THE ART OF PROBABILITY
FOR SCIENTISTS AND ENGINEERS

The Art of Probability

FOR SCIENTISTS AND ENGINEERS

Richard W. Hamming

U. S. Naval Postgraduate School

Hamming, R. W. (Richard Wesley)

Addison-Wesley Publishing Company, Inc.
The Advanced Book Program
Redwood City, California • Menlo Park, California
Reading, Massachusetts • New York • Don Mills, Ontario
Wokingham, United Kingdom • Amsterdam • Bonn
Sydney • Singapore • Tokyo • Madrid • San Juan

QA
273
.43544
1991

Publisher: *Allan M. Wylde*
Marketing Manager: *Laura Likely*
Production Manager: *Jan V. Benes*
Production Assistant: *Karl Matsumoto*
Electronic Composition and Text Design: *Peter Vacek*
Cover Design: *Iva Frank*

Library of Congress Cataloging-in-Publication Data

Hamming, R. W. (Richard Wesley), 1915–
    The art of probability–for scientists and engineers/Richard W. Hamming.
        p.   cm.
    Includes index.
    1. Probabilities.   I. Title.
    QA273.H3544   1991
    519.2–dc20   90-42240
    ISBN 0-201-51058-8   CIP

This book was typeset using the TEX typesetting language on IBM Compatible computer and output on HP LaserJet Series II.

2345678910 –MA– 95 94 93 92 91

Preface

"If we want to start new things rather than trying to elaborate and improve old ones, then we cannot escape reflecting on our basic conceptions."   Hans Primas [P, p. 17]

Every field of knowledge has its subject matter and its methods, along with a style for handling them. The field of Probability has a great deal of the Art component in it—not only is the subject matter rather different from that of other fields, but at present the techniques are not well organized into systematic methods. As a result each problem has to be "looked at in the right way" to make it easy to solve. Thus in probability theory there is a great deal of art in setting up the model, in solving the problem, and in applying the results back to the real world actions that will follow. It is necessary to include some of this art in any textbook that tries to prepare the reader to *use* probability in the real world of science and engineering rather than merely admire it as an abstract discipline and a branch of mathematics.

It is widely agreed that art is best taught through concrete examples. Especially in teaching probability it is necessary to work many problems. Since the answers are already known the purpose of the Examples and Exercises cannot be to "get the answer" but to illustrate the *methods* and *style* of thinking. Hence the Examples in the text should be studied for the methods and style as well as for the results; also they often have educational value. Thus in solving the Exercises style should be considered as part of their purpose.

It is not enough to merely give solutions to problems in probability. If the art is to be communicated to the reader then the initial approach—which is so vital in this field—must be carefully discussed. From where, for example, do the initial probabilities come? Only in this way can the reader learn this art—and most mathematically oriented text book simply ignore the source of the probabilities!

What is probability? I asked myself this question many years ago, and found that various authors gave different answers. I found that there were several main schools of thought with many variations. First, there were the frequentists who believe that probability is the limiting ratio of the successes divided by the total number of trials (each time repeating *essentially* the same situation). Since I have a scientific-engineering background, this approach, when examined in detail, seemed to me to be non-operational and furthermore excluded important and interesting situations.

Second, there are those who think that there is a probability to be attached to a single, unique event without regard to repetitions. Via the law of large numbers they deduce the frequency approach as something that is likely, but not sure.

Third, I found, for example, that the highly respected probabilist di Finetti [dF, p. x] wrote at the opening of his two volume treatise,

Probability Does Not Exist.

Fourth, I found that mathematicians tend to simply postulate a Borel family of sets with suitable properties (often called a σ-algebra or a Borel field) for the sample space of events and assign a measure over the field which is the corresponding probability. But the main problems in using probability theory are the choice of the sample space of events and the assigment of probability to the events! Some highly respected authors like Feller [F, p. x] and Kac [K, p. 24] were opposed to this measure theoretic mathematical approach to probability; both loudly proclaimed that probability is not a branch of measure theory, yet both in their turn seemed to me to adopt a formal mathematical approach, as if probability were merely a branch of mathematics and not an independent field.

Fifth, I also found that there is a large assortment of personal probabilists, the most prominent being the Bayesians, at least in volume of noise. Just what the various kinds of Bayesians are proclaiming is not always clear to me, and at least one said that while nothing new and testable is produced still it is "the proper way to think about probability".

Finally, there were some authors who were very subjective about probability, seeming to say that each person had their own probabilities and there need be little relationship between their beliefs; possibly true in some situations but hardly scientific.

When I looked at the early history of probability I found the seeds of most of these views; apparently very little has been settled in all these years.

I also found that there were flagrant omissions in all the books; whole areas of current use of probability, such as quantum mechanics, were completely ignored! Few authors cared to even mention them.

What was I to make of all this? Following an observation of Disraeli, I decided to find out what I myself believed by the simple process of writing a book that would be, to me at least, somewhat more believable than what had I found. Of necessity, being application oriented, it would have a lot more philosophy than most textbooks that postulate a probability model and then present the techniques without regard to understanding when, how, and where to use the techniques, or why the particular postulates are assumed. Indeed, most mathematicians blandly *assume* that probability is a branch of mathematics without ever doubting this assumption. Such books tend to discourage the reader from making new applications outside the accepted areas—yet it is hard to believe that the range of applications of probability has been anywhere near exhausted. It seems to me that the philosophy of probability is not a topic to be avoided in a first course, but rather, in view of the dangers of the misapplication of the theory (which are many), it is an essential part.

Initially I wanted to provide some organization and structure for the various methods for solving probability problems instead of merely giving the usual presentation of problems and their solution by any method, selected almost at random, that would work. My success has been of limited extent, but I feel that I have taken a few steps in that direction.

I further decided that it would be necessary to build up my intuition about problems so that: (1) many of the false results that are so easily obtained would be noticed, and (2) that even if I could not solve a problem still I might have a feeling for the size and nature of the answer.

Finally, as a sometime engineer, I well know that few things are known exactly, and that many times probabilities used in the final result are based on estimates, hence the *robustness* (sensitivity) of the results with respect to the small changes in the initially assumed probabilites and their interrelationships must be investigated carefully. Again, this is a much neglected part of probability theory, though clearly it is essential for serious applications.

The availability of computers, even programmable hand held ones, greatly affects probability theory. First, one can easily evaluate formulas that would have taxed hand computation some decades ago. Second, often the *simulation* of probability problems is now very practical [see Chapter 10]; not that one can get answers accurate to many decimal places, but that a crude simulation can reveal a missed factor of 2, the wrong sign on a term, and other gross errors in the formula that purports to be the answer. These simulations can also provide some intuition as to why the result is what it is, and even at times suggest how to solve the problem analytically. Simulation can, of course, give insight to problems we cannot otherwise solve.

The fact that probability theory is increasingly being used to make important decisions is a further incentive for examining the theory carefully. I have witnessed very important decisions being made that were based on probability, and as a citizen I have to endure the consequences of similar decisions made in Washington D.C. and elsewhere; hence I feel that anything that can clarify and improve the quality of the application of probability theory will be of great benefit to our society.

It is generally recognized that it is dangerous to apply any part of science without understanding what is behind the theory. This is especially true in the field of probability since in practice there is not a single agreed upon model of probability, but rather there are many widely different models of varying degrees of relevance and reliability. Thus the philosophy behind probability should not be neglected by presenting a nice set of postulates and then going forward; even the simplest applications of probability can involve the underlying philosophy. The frequently made claim that while the various foundational philosophies of probability may be different, still the subsequent technique (formalism) is always the same, *is flagrantly false!* This book gives numerous examples illustrating this fact. One can only wonder why people make this claim; the reason is probably the desire to escape the hard thinking on the foundations and to get to protection of the formalism of mathematics. Furthermore, the interpretation of the results may be quite difficult and require careful thinking about the underlying model of probability that was assumed.

This book is the result. It is one man's opinion using a rather more scientific (as opposed to mathematical) approach to probability than is usual. It is hardly perfect, leaves a lot open for further work, and omits most of subjective probability as not being scientific enough to justify many actions in the real world based on it. Not only are many scientific theories and problems based on probability, but many engineering tasks, such as the launching of space vehicles, depend on probability estimates. Perhaps most important, many political, biological, medical, and social decisions involve probability in an essential way. One would like to believe that these decisions are based on sound principles and not on personal prejudices, politics and propaganda.

In order to strengthen the reader's intuitions for making probability judgements, I have included a reasonable number of tables of results. These tables are well worth careful study to understand why the numbers are the way they are. I have also examined various results to show how they agree, or disagree, with common experience. In my opinion this is a necessary part of any course in probability, since in normal living one has a very limited exposure to the varieties of peculiar results that can occur.

The material has therefore been carefully presented in a pedagogical manner, including deliberate repetitions, for the benefit of the beginner, and not in the logical order for the benefit of the professor who already understands probability.

But one's efforts are limited, and it occurred to me that others might want to examine, criticize, change, and advance further the problem of what probability is, hence what I found is presented here for their consideration. I doubt that in the future there will be any single, widely accepted model of probability that is useful for all applications; hence the need for multiple approaches to the topic, and in time new ones not yet discovered!

I am greatly indebted to Professor Roger Pinkham of Stevens Institute,

Hoboken, for endless patience and guidance while I tried to learn probability theory, as well as for a large supply of Examples to illustrate various points. I am, of course, solely responsible for the contents of this book and he cannot be held responsible for my opinions and errors; he did his best!

I am also indebted to Professor Bruce MacLennan of the University of Tennesee, Knoxville, for many stylistic improvements and suggestions for a clearer presentation; again he is not responsible for the final product. Professor Don Gaver has been of help in numerous discussions.

Table of Contents

1

Probability

"Probability is too important to be left to the experts."

1.1 Introduction

Who has not watched the toss of a coin, the roll of dice, or the draw of a card from a well shuffled deck? In each case, although the initial conditions appear to be much the same, the specific outcome is not knowable. These situations, and many equivalent ones, occur constantly in our society, and we need a theory to enable us to deal with them on a rational, effective basis. This theory is known as *probability theory* and is very widely applied in our society—in science, in engineering, and in government, as well as in sociology, medicine, politics, ecology, and economics.

How can there be laws of probability? Is it not a contradiction? Laws imply regularity, while probable events imply irregularity. We shall see that many times amid apparent irregularity some regularity can be found, and furthermore this limited regularity often has significant consequences.

How is it that from initial ignorance we can later deduce knowledge? How can it be that although we state that we know nothing about a single toss of a coin yet we make definite statements about the result of many tosses, results that are often closely realized in practice?

Probability theory provides a way, indeed a *style,* of thinking about such problems and situations. The classical theory has proved to be very useful even in domains that are far removed from gambling (which is where it arose and is still, probably, the best initial approach). *This style of thinking is an art* and is not easy to master; both the historical evidence based on its late development, and the experience of current teaching, show that much *careful thinking on the student's part* is necessary before probability becomes a mental habit. The sophisticated approach of beginning with abstract postulates is favored by mathematicians who are interested in covering as rapidly as possible the material and techniques that have been developed in the past. This is not an effective way of teaching the understanding and the use of probability in new situations, though it clearly accelerates the formal manipulation of the symbols (see the quotation at the top of the Preface). We adopt the slow, cautious approach of carefully introducing the assumptions of the models, and then examining them, together with some of their consequences, before plunging into the formal development of the corresponding theory. We also show the relationship between the various models of probability that are of use in practice; there is not a single model of probability, but many and of differing reliabilities.

Mathematics is not just a collection of results, often called theorems; it is a style of thinking. Computing is also basically a style of thinking. Similarly, probability is a style of thinking. And each field is different from the others. For example, I doubt that mathematics can be reduced to button pushing on a computer and still retain its style of thinking. Similarly, I doubt that the theory of probability can be reduced to either mathematics or computing, though there are people who claim to have done so.

In normal life we use the word "probable" in many different ways. We speak of the probability of a head turning up on the toss of a coin, the probability that the next item coming down a production line will be faulty, the probability that it will rain tomorrow, the probability that someone is telling a lie, the probability of dying from some specific disease, and even the probability that some theory, say evolution, special relativity, or the "big bang," is correct.

In science (and engineering) we also use probability in many ways. In its early days science assumed there was an exact measurement of something, but there was also some small amount of "noise" which contaminated the signal and prevented us from getting the exact measurement. The noise was modeled by probability. We now have theories, such as *information theory* and *coding theory*, which assume from the beginning that there is an irreducible noise in the system. These theories are designed to take account of the noise rather than to initially avoid it and then later, at the last moment, graft on noise. And there are some theories, such as *the Copenhagen interpretation of quantum mechanics*, which say that probability lies at the foundations of physics, that it is basic to the very nature of the world.

Thus there are many different kinds of probability, and any attempt to give only one model of probability will not meet today's needs, let alone tomorrow's. Some probability theories are "subjective" and have a large degree of "personal probability" (belief) as an essential part. Some theories try to be more scientific (objective), and that is the path we will mainly follow— without prejudice to the personal (more subjective) probability approaches which we cannot escape using in our private lives. [Kr, both volumes].

The belief that there is not a single model of probability is rarely held, but it is not unique to the author. The following paraphrase shows an alternate opinion (ignore the jargon for the moment) [G, p.xi,xii]:

> In quantum probability theory the state of the system is determined by a complex-valued function A (called the amplitude function) on the outcome space (or sample space). The outcomes of an event $E = \{x_1, x_2, \ldots\}$ have the probability $P(E)$ of E which is computed by
>
> $$P(E) = |\sum [A(x_i)]|^2$$
>
> This allows for interference as happens in optics and sound. If the outcomes do not interfere, then of course
>
> $$P(E) = \sum [|A(x_i)|]$$
>
> This last is the classical probability theory. In the interference case $P(E)$ decomposes into the sum of two parts. The real part is the classical counterpart and the imaginary part leads to the constructive or destructive interference which is characteristic of quantum mechanics and much of physics.
>
> The basic axiom of the path integral formalism is that the state of a quantum mechanical system is determined by the amplitude function A and that the probability of a set of interfering outcomes $\{x_1, x_2, \ldots\}$ is the first equation. This point has been missed by the axiomatic approaches and in this sense they have missed the essence of quantum mechanics. In principle this essence can be regained by adjoining an amplitude function axiom to the other axioms of the system, but then the axiomatic system is stronger than necessary and may well exclude important cases.

Since we are going to use a scientific approach to probability it is necessary to say what science is and is not. In the past science has tried to be objective and to insist on the property that different people doing the same experiment would get essentially the same results. But we must not assume that science is completely objective, that this repeatability property is perfectly attained or is even essential (for example, consider carefully observational astronomy). Again, if I were to ask two different people to measure the width of a table, one of them might by mistake measure the length, and hence the measurements might be quite different. There is an unstated amount of "understanding what is meant" that is implied in every experimental description. We can only hope to control this assumed background (that is supposed to produce absolute consistency) so that there is little room for error, but we do not believe that we can ever completely eliminate some personal judgment in science. In science we try to identify clearly where judgment comes in and to control the amount of it as best we can.

The statement at the head of this chapter,

Probability is too important to be left to the experts,

is a particular case of a very general observation that the experts, by their very expert training and practice, often miss the obvious and distort reality seriously. The classical statement of this general principle is, "War is too important to be left to the generals." We daily see doctors who treat cases but not patients, mathematicians who give us exact solutions to the wrong problems, and statisticians who give us verifiably wrong predictions. In our Anglo-Saxon legal system we have clearly adopted the principle that in guilt vs. innocence "twelve tried and true men" are preferable to an experienced judge. The author clearly believes that the same is true in the use of probability; the desire of the experts to publish and gain credit in the eyes of their peers has distorted the development of probability theory from the needs of the average user.

The comparatively late rise of the theory of probability shows how hard it is to grasp, and the many paradoxes show clearly that we, as humans, lack a well grounded intuition in this matter. Neither the intuition of the man in the street, nor the sophisticated results of the experts provides a safe basis for important actions in the world we live in. The failure to produce a probability theory that fits the real world as it is, more or less, can be very costly to our society which uses it constantly.

In the past science has developed mainly from attempts to "measure how much," to be more precise than general statements such as larger, heavier, faster, more likely, etc. Probability tries to measure more precisely than "more likely" or "less likely." In order to use the scientific approach we will start (Section 1.5) with the *measurement* of probability in an objective way. But first some detours and formalities are needed.

1.2   Models in General

It is reasonably evident to most people that all thought occurs internally, and it is widely believed that thinking occurs in the head. We receive stimuli from the external world, (although some sophists have been known to deny the existence of the external world, we shall adopt a form of the "naive realism" position that the world exists and is to some extent "knowable"). We organize our incoming stimuli according to some ill-defined model of reality that we have in our heads. As a result we do not actually think about the real external world but rather we "think" in terms of our impressions of the stimuli. Thus *all* thought is, in some sense, *modeling*.

A model can not be proved to be correct; at best it can only be found to be reasonably consistant and not to contradict some of our beliefs of what reality is. Karl Popper has popularized the idea (which goes back at least to Francis Bacon) that for a theory to be scientific it must be potentially *falsifiable*, but in practice we do not always obey his criterion, plausible as it may sound at first. From Popper's point of view a scientific theory can only be disproved, but never proved. Supporting evidence, through the application of finite induction, can increase our faith in a theory but can never *prove* its absolute truth.

In actual practice many other aspects of a model are involved. Occam's razor, which requires that minimal assumptions be used and extra ones removed, is often invoked as a criterion—we do not want redundant, possibly contradictory assumptions. Yet Occam's razor is not an absolute, final guide. Another criterion is æsthetic taste, and that depends, among other things, on the particular age and culture you happen to live in. We feel that some theories are nicer than others, that some are beautiful and some are ugly, and we tend to choose the loveliest one in practice. Another very important criterion is "fruitfulness"—does the model suggest many new things to try?

A model is often judged by how well it "explains" some observations. There need not be a unique model for a particular situation, nor need a model cover every possible special case. A model is not reality, it merely helps to explain some of our impressions of reality. For example, for many purposes we assume that a table is "solid," yet in physics we often assume that it is made of molecules and is mainly "empty space." Different models may thus seem to contradict each other, yet we may use both in their appropriate places.

In view of the wide variety of applications of probability, past, present, and future, we will develop a sequence of models of probability, generally progressing from the simpler to the more complex as we assume more and more about the model. At each stage we will give a careful discussion of the assumptions, and illustrate some features of them by giving consequences you are not likely to have thought were included in the model. Thus we will regularly discuss examples (paradoxes) whose purpose is to show that you can get rather peculiar results from a model if you are not careful, even though the assumptions (postulates) on the surface seem to be quite bland, [St]. It is irresponsible to teach a probability course as if the contents were entirely safe to use in any situation. As Abelard (1079–1142) said,

> *"I expose these contradictions so that they may excite the susceptible minds of my readers to the search for truth, and that these contradictions may render their minds more penetrating as the effect of that search."*

Since the applications of probability theory are constantly expanding in range and depth, it is unlikely that anyone can supply models for all the situations that will arise in the future. Hence we are reduced to presenting a series of models, and to discussing some of their relationships, in the hopes that you will be able to create the model you need when the time comes to develop a new one. Each application you make requires *you* to decide which probability model to adopt.

1.3   The Frequency Approach Rejected

Most people have at least two different models of probability in their minds. One model is that there is a probability to be associated with a single event, the other is that probability is the ratio of the number of successes to the total number of trials in a potentially infinite sequence of similar trials. We need to assume one of these and deduce, in so far as we can, the other from the first—otherwise we risk assuming a contradiction! It will turn out that *the two models are not completely equivalent*.

A third kind of probability that most people are familiar with is that which arises in betting between friends on sporting events. The "odds" are simply a version of probability; the odds of a to b are the probabilities of $a/(a+b)$ and $b/(a+b)$. There is often no serious thought of repetitions of the event, or a "random selection from an ensemble of similar events," rather it is the feeling that your "hunches" or "insights" are better than those of the opponent. Both sides recognize the *subjective* aspect, that two people with apparently the same information can have different probability assignments for the same event. The argument that they have different amounts of information is unconvincing; it is often merely a temporary mood that produces the difference.

The *frequency approach* (the ratio of the number of successes to the total number of equally likely possible outcomes) seems to be favored by most statistically inclined people, but it has severe scientific difficulties. The main difficulty with the frequency approach is that no one can say how many (necessarily finite) number of trials to make, and it is not practical science to define the probability as the limit of the ratio as the number of trials approaches infinity. Furthermore, the statement of the limiting frequency, in a careful form, will often include the words that you will for any long, finite sequence be *only probably* close! It is circular to say the least! And you can't *know* whether you are looking at an exceptional case or a likely case. It is also true that to get even reasonable accuracy usually requires a discouragingly large number of trials. Finally, there are models, as we shall show in Section 8.7 for which the single event probability has meaning and at the same time the average of n samples has more variability than the original distribution!

For a long time there has been in science a strong trend towards accepting in our theories only objectively measureable quantities and of excluding purely conceptual ones that are only props (like the "ether" in classical physics). The extreme of allowing *only* measureable quantities is perhaps too rigid an attitude (though a whole school of philosophy maintains this position); yet it is a useful criterion in doing science.

Similarly, there are many mathematical theorems in the literature whose hypotheses cannot be known, even in an intuitive sense, to be close let alone be either exactly true or false within the framework of the intended application; hence these theorems are not likely to be very useful in these practical applications of the theory. Such theorems may be interesting and beautiful pure mathematics but are of dubious applicability.

This "practical use" attitude tends to push one towards "constructive mathematics" [B] and away from the conventional mathematics. The increasing use of computers tends to make many people favor "computable" numbers as a basis for the development of probability theory. Without careful attention to the problem for which it is being used it is not a priori obvious which approach to the real number system we should adopt. And there are some people who regard the continuous mathematics as merely a useful approximation to the discrete since usually the equivalent operations are easier in the continuous model than they are in the discrete. Furthermore, quantum mechanics suggests, at least to some people, that ultimately the universe is discrete rather than continuous.

Whenever a new field of application of mathematics arises it is always an open question as to the applicability of the mathematics which was developed earlier; if we are to avoid serious blunders it must be carefully rethought. This point will be raised repeatedly in this book.

For these and other reasons we will not begin with the frequency approach, but rather deduce (Sections 2.8 and Appendix 2.A) the frequency model along with its limitations. For this present Section see [Ke], especially Chapter 8.

1.4 The Single Event Model

In most fields of knowledge it is necessary, sooner or later, to introduce some technical notation and jargon so that what is being talked about is fairly precise and we can get rid of the vagueness of words as used in normal language. We have used some of these words already!

At first a *trial* is a single unit, like the toss of a coin, the roll of a die (singular of dice), or the draw of a single card from a well shuffled deck. The *elementary events* of a trial are the possible outcomes of the trial. For example, for the toss of a coin the elementary events are "head" and "tail."

For the roll of a die the elementary events are the names of the six faces, $1, 2, 3, 4, 5, 6$. The elementary events are considered to be indivisible, atomic, events from which more complex events (say an even numbered face of a die) may be built. We will label the possible outcomes of a trial by the symbols x_i, where i runs from 1 to n (or possibly 0 to n). Thus as a first approximation, to be modified later, we have the trial labeled X whose outcome is some particular x_i. The capital letter X represents the trial, and the set $\{x_i\}$ represents the possible outcomes of the trial; some particular x_i will be the actual outcome of the trial X being conducted. In some circumstances the symbol X may be thought of as being the whole set $\{x_i\}$ just as the function $\sin x$ may be thought of as the entire function. This list of possible outcomes is an idealization of reality. Initially we assume that this list is finite in length.

Later on a trial may be a sequence of these simple trials; the roll of, say, four dice at one time will be a single trial. Still later we will continue for an indefinite number of times until some condition is reached such as tossing a coin until three heads occur in a row. In such cases the entire run will be considered a single trial, and the number of possible events will not be finite but rather countably infinite. Still later, we will examine a continuum of events, a classically non-countable number of events.

In principle the first step in a discrete probability problem is, for the trial (named) X, to get the corresponding list (or set) X of all the possible outcomes x_i. Thus X has two meanings; it is the name of the trial and it is also the list of possible the outcomes of the trial. The name of the outcome should not be logically confused with any *value* associated with the outcome. Thus logically the face "the number of spots," say 4, on a die is a label and should not be confused with the name "four," nor with the value "4" that is usually associated with this outcome. In practice these are often carelessly treated as being the same thing.

The beginner is often confused between the model and reality. When tossing a coin the model will generally not include the result that the coin ends up on its edge, nor that it is lost when it rolls off the table, nor that the experimenter drops dead in the middle of the trial. We could allow for such outcomes if we wished, but they are generally excluded from simple problems.

In a typical mathematical fashion these events (outcomes) x_i are considered to be points in the *sample space* X; each elementary event x_i is a point in the (at present) finite, discrete sample space. To get a particular realization of a trial X we "sample" the sample space and select "at random" (to be explained in more detail in Section 1.9) an event x_i. The space is also called a *list* (as was stated above) and we choose (at random) one member (some x_i) of the list X as the outcome of the trial.

1.5 Symmetry as the Measure of Probability

With this notation we now turn to the assignment of a *measure* (the numerical value) of the *probability* p_i to be associated with (assigned to) the event x_i of the trial.

Our first tool is *symmetry*. For the *ideal* die that we have in our mind we see that (except for the labels) the six possible faces are all *the same*, that they are *equivalent*, that they are *symmetric*, that they are *interchangeable*. Only one of the elementary events x_i can occur on one (elementary) trial and it will have the probability p_i. If any one of several different outcomes x_i is allowed on a trial (say either a 4 or a 6 on a roll of a die) then we must have the "additivity" of probabilities p_i. For the original sample space we will get the same probability, due to the assumed symmetry, for the occurences of each of the elementary events. We will scale this probability measure so that the sum of all the probabilities over the sample space of the elementary events must add up to 1.

If you do not make equal probability assignments to the interchangeable outcomes then you will be embarrassed by interchanging a pair of the nonequal values. Since the six faces of an ideal die are all symmetric, interchangeable if you prefer, then each face must have the probability of 1/6 (so that the total probability is 1). For a well shuffled *ideal* deck of cards we believe that each of the 52 cards are interchangeable with every other card so each card must have a probability of 1/52. For a well balanced coin having two symmetric faces each side must have probability 1/2.

Let us stop and examine this crucial step of assigning a measure (number) p_i called "the probability" to each possible outcome x_i of a trial X. These numbers are the probabilities of the elementary events. From the perceived symmetry we deduced interchangeability, hence the equality of the measure of probability of each of the possible events. The reasonableness of this clearly reinforces our belief in this approach to assigning a measure of probability to elementary events in the sample space.

This perception of symmetry is subjective; either you see that it is there or you do not. But at least we have isolated any differences that may arise so that the difference can be debated sensibly. We have proceeded in as objective a fashion as we can in this matter.

However, we should first note that, "Symmetric problems always have symmetric solutions." is false [W]. Second, when we say symmetric, equivalent, or interchangable, is it based on perfect knowledge of all the relevant facts, or is it merely saying that we are not aware of any lack thereof? The first case being clearly impossible, we must settle for the second; but since the second case is vague as to how much we have considered, this leaves a lot to be desired. We will always have to admit that if we examined things more closely we might find reason to doubt the assumed symmetry.

We now have two general rules: (1) *if there are exactly n different inter-*

changeable (symmetric in some sense) elementary events x_i, then each event x_i has probability $p_i = 1/n$ since (2) the probabilities of the elementary events in the sample space are additive with a total probability of 1.

Again, from the additivity of the probabilities of the elementary events it follows that for more complex situations when there is more than one point in the sample space that satisfies the condition then we add the individual probabilities.

For the moment we will use the words "at random" to mean equally likely choices in the sample space; but even a beginner can see that this is circular— what can "equally likely" mean except equally probable, (which is what we are trying to define)? Section 1.9 will discuss the matter more carefully; for the moment the intuitive notion will suffice to do a few simple problems.

Example 1.5–1   *Selected Faces of a Die*

Suppose we have a well balanced die and ask for the probability that on a random throw the upper face is less than or equal to 4.

The admissible set of events (outcomes) is $\{x_i\} = \{1, 2, 3, 4\}$. Since each has probability $p_i = 1/6$, we have

$$\text{Prob } \{x_i \leq 4\} = 1/6 + 1/6 + 1/6 + 1/6 = 4/6 = 2/3$$

If instead we ask for the probability that the face is an even number, then we have the acceptable set $\{2, 4, 6\}$, and hence

$$\text{Prob } \{x_i \text{ even}\} = 1/6 + 1/6 + 1/6 = 3/6 = 1/2$$

Example 1.5–2   *A Card With the Face 7*

What is the probability that on the random draw of a card from the standard card deck of 52 cards (of 4 suits of 13 cards each) that the card drawn has the number 7 on it?

We argue that there are exactly 4 cards with a 7 on their face, one card from each of the four suits, hence since the probability of a single card is 1/52 we must have the probability

$$\text{Prob } \{\text{face} = 7\} = 4/52 = 1/13.$$

There are other simple arguments that will lead to the same result.

Example 1.5–3   *Probability of a Spade*

What is the probability of drawing a spade on a random draw from a deck of cards?

We observe that the probability of any card drawn at random is 1/52. Since there are 13 spades in the deck we have the combined probability of

$$P(\text{spade}) = (13)(1/52) = 1/4$$

Exercises 1.5

1.5–1. If there are 26 chips each labeled with a different letter of the alphabet what is the probability of choosing the chip labeled Q when a chip is drawn at random? Of drawing a vowel (A, E, I, O, U)? Of not drawing a vowel? Ans. 1/26, 5/26, 21/26.

1.5–2. If 49 chips are labeled $1, 2, \ldots 49$, what is the probability of drawing the 7 at random? A number greater or equal to 7? Ans. 1/49, 43/49.

1.5–3. If 49 chips are labeled $2, 4, \ldots 98$, what is the probability of drawing the chip 14 at random? Of a number greater than 20?

1.5–4 A plane lattice of points in the shape of a square of size $n > 2$ has n^2 points. What is the probability of picking a corner point? The probability of picking an inner point? Ans. $4/n^2, [(n-2)/n]^2$

1.5–5 A day of the week is chosen at random, what is the probability that it is Tuesday? A "weekday"?

1.5–6 We have n^3 items arranged in the form of a lattice in a cube, and we pick at random an item from an edge (not on the bottom). What is the probability that it is a corner item? Ans. $1/(2n-3)$.

1.5–7 In a set of 100 chips 10 are red; what is the probability of picking a red chip at random? Ans. 1/10.

1.5–8 If there are r red balls, w white balls, and b blue balls in an urn and a ball is drawn at random, what is the probability that the ball is red? Is white? Is blue? Ans. $r/(r+w+b), w/(r+w+b), b/(r+w+b)$.

1.5–9 There are 38 similar holes in a roulette wheel. What is the probability of the ball falling in any given hole?

1.5–10 In a deck of 52 cards what is the probability of a random drawing being the 1 (ace) of spades? Of drawing at random a heart? Ans. 1/52, 1/4

1.5–11 There are 7 bus routes stopping at a given point in the center of town. If you pick a bus at random what is the probability that it will be the one to your house?

1.5–12 In a box of 100 items there are 4 defective ones. If you pick one at random what is the probability that it is defective? Ans. 1/25

1.5–13 If 3% of a population has a certain disease what is the probability that the next person you pass on the street has it? Ans. 0.03

1.5–14 Due to a maladjustment a certain machine produces every fifth item defective. What is the probability that an item taken at random is defective?

1.5–15 In a certain class of 47 students there are 7 with last names beginning with the letter S. What is the probability that a random student's name will begin with the letter S? Ans. 7/47

1.6 Independence

The next idea we need is *independence*. If we toss a coin and then roll a die we feel that the two outcomes are *independent*, that the outcome of the coin has no influence on the outcome of the die. The sample space is now

	1	2	3	4	5	6
H	H1	H2	H3	H4	H5	H6
T	T1	T2	T3	T4	T5	T6

We regard this sample space as the *product* of the two elemetary sample spaces. This is often called the Cartesian (or direct) product by analogy with cartesian coordinates of the two original sample spaces, $\{H, T\}$ for the coin, and $\{1, 2, 3, 4, 5, 6\}$ for the die. Since we believe that the outcomes of the die and coin are independent we believe (from symmetry) that the events in the product sample space are all equally likely, hence each has probability $1/(2 \times 6) = 1/12$. We see that the probability of the compound events in the product sample space are (for independent events) just the product of the corresponding probabilities of the separate original events, that

$$1/12 = (1/2)(1/6).$$

Clearly the product space (meaning the product of the sample spaces) of the coin with the die is the same as the product space of the die with the coin.

We immediately *generalize* to any two independent events, each of these being from an equally likely sample space. If there are n_1 of the first kind of event and n_2 of the second, then the product space of the combined trial has $n_1 n_2$ events each of the same probability $1/n_1 n_2 = (1/n_1)(1/n_2)$.

Example 1.6–1   *The Sample Space of Two Dice*

Consider the toss of two independently rolled dice. The sample space has $6 \times 6 = 36$ events each of probability 1/36, and the sample space is:

$$
\begin{array}{ccccc}
1,1 & 1,2 & 1,3 & \ldots & 1,6 \\
2,1 & 2,2 & 2,3 & \ldots & 2,6 \\
\ldots & \ldots & \ldots & \ldots & \ldots \\
6,1 & 6,2 & 6,3 & \ldots & 6,6
\end{array}
$$

Clearly the idea of a product space may be *extended* to more than two independent things.

Example 1.6–2   *The Sample Space of Three Coins*

Consider the toss of three coins. We have $2 \times 2 \times 2 = 8$ equally likely outcomes. It is awkward to draw 3 dimensional sample spaces, and it is not feasible to draw higher dimensional sample spaces, so we resort to listing the elementary events. For the three coins we have a list (equivalent to the product sample space)

$$HHH \quad HHT \quad HTH \quad THH \quad HTT \quad THT \quad TTH \quad TTT$$

each of probability $(1/2)(1/2)(1/2) = 1/8$.

The extension to m different independent trials each having $n_i(i = 1, 2, \ldots, m)$ equally likely outcomes leads to a product space of size equal to the product $n_1 n_2 \ldots n_m$ of the dimensions of the original spaces. The probabilities of the events in the product space are equal to the corresponding products of the probabilities of the original events making up the original spaces, hence the product space events are also equally likely (since the original ones were equally likely).

Example 1.6–3   *The Number of Paths*

If you can go from A to B in 6 ways, from B to C in 7 ways, and from C to D in 5 ways, then there are $6 \times 7 \times 5 = 210$ ways of going from A to D, Figure 1.6–1. If each of the original ways were equally likely and independent then each of the 210 ways is also equally likely and has a probability of 1/210.

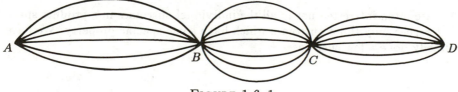

FIGURE 1.6–1

Notice again, the probabilities of the equally likely events in the product sample space can be found by either: (1) computing the number of events x_i in the product space and assigning the same probability $p_i = 1/($the number of events$)$ to each event in the product space, or (2) assign the product of the probabilities of the separate parts to each compound event in the product space.

Notice also that we are always *assuming* that the elementary points in the sample space are independent of each other, that one outcome is not influenced by earlier choices. However, in this universe in which we live *apparently* all things are interrelated ("intertwined" is the current jargon in quantum mechanics); if so then independence is an idealization of reality, and is not reality.

It is easy to make up simple examples illustrating this kind of probability problem, and working out a large number of them is not likely to teach you as much as carefully reviewing in your mind *why* they work out as they do.

Exercises 1.6

1.6–1 What is the probability of HHH on the toss of three coins when the order of the outcomes is important? Of HTH? Of TTH? Ans. 1/8.

1.6–2 Describe the product space of the draw of a card and the roll of a die.

1.6–3 What is the size of the product space of the roll of three dice? Ans. $6^3 = 216$.

1.6–4 What is the size of the product space of the toss of n coins? Ans. 2^n.

1.6–5 What is the size of the product space of the toss of n dice?

1.6–6 What is the product space of a draw from each of two decks of cards? Ans. Size= $(52)^2$.

1.6–7 What is the size of the sample space of independent toss of a coin, roll of a die, and the draw of a card?

1.7   Subsets of a Sample Space

We are often interested not in all the individual outcomes in a product space but rather only those outcomes which have some specific property. We have already done a few such cases.

Example 1.7–1   *Exactly One Head in Three Tosses*

Suppose we ask in how many ways can exactly one head turn up in the toss of three coins, or equivalently, what is the probability of the event. We see that in the complete sample space of 8 possible outcomes we are only interested in the three events

$$HTT \quad THT \quad TTH$$

Each of these has a probability of 1/8, so that the sum of these will be 3/8 which is the corresponding probability.

Clearly we are using the additivity of the probabilities of the events in the sample space. We need to know the size of the sample space, but we need not list the whole original sample space, we need only count the number of successes. Listing things is merely one way of counting them.

Example 1.7–2   *Sum of Two Dice*

What is the probability that the sum of the faces of two independently thrown dice total 7? We have only the six successful possibilities

$$1+6, \quad 2+5, \quad 3+4, \quad 4+3, \quad 5+2, \quad 6+1$$

in the total sample space of $6^2 = 36$ possible outcomes. See Figure 1.7-1. Each possible outcome has probability of 1/36, hence the probability of a total of 7 is $6/36 = 1/6$.

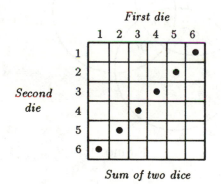

First die

Second die

Sum of two dice

FIGURE 1.7–1

We can look at this in another way. Regardless of what turns up on the first die, there is exactly one face of the second that will make the total equal to 7, hence the probability of this unique face is 1/6.

If we were to ask in how many ways, (or equivalently what is the probability), can the sum of three or four, or more dice add up to a given number then we would have a good deal of trouble in constructing the complete sample space and of picking out the cases which are of interest. It is the purpose of the next two chapters to develop some of the mathematical tools to carry out this work; the purpose of the present chapter is to introduce the main ideas of probability theory and not to confuse you by technical details, so we will not now pursue the matter much further.

We now have the simple rule: count the number of different ways the compound event can occur in the original sample space of equally likely events (outcomes), then the probability of this compound event is the ratio of the number of successes in the sample space to the total number of events in the original sample space.

Aside: *Probability Scales*

We have evidently chosen to measure probability on a scale from 0 to 1, with 0 being impossibility and 1 being certainty. A scale that might occur to you is from 1 to infinity, which is merely the reciprocal of the first range, (you might also consider 0 to infinity). In these cases you will find that the addition of probabilities is not simple. There are other objections that you can find if you choose to explore these scales.

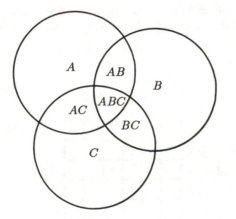

Venn Diagram

FIGURE 1.7–2

A *Venn diagram* is mainly a symbolic representation of the subsets of a sample space. Typically circles are used to represent the subsets, see Figure 1.7-2. The subsets represented are the insides of the circles A, B and C. Now consider those events which have both properties A and B, shown by the region AB. Similarly for the sets AC and BC. Finally, consider the events which have the properties of A, B and C at the same time; this is the inner piece in the diagram, ABC. In this simple case the representation is very illuminating. But if you try to go to very many subsets then the problem of drawing a diagram which will show clearly what you are doing is often difficult; circles are not, of course, necessary, but when you are forced to draw very snake-like regions then the diagram is of little help in visualizing the situation. In the Venn diagram the members of a set are often scattered around (in any natural display of the elements of the sample space), hence we will not use Venn diagrams in this book.

Exercises 1.7

1.7–1 What is the probability of a die showing an odd digit?

1.7–2 What is the probability of drawing a "diamond" when drawing a card from a deck of 52 cards? Ans. 1/4.

1.7–3 On the roll of two dice what is the probability of getting at least one face as a 4? Ans. 11/36.

1.7–4 What is the probability of two H's in two tosses of a coin?

1.7–5 On the roll of two independent dice what is the probability of a total of 6? Ans. 5/36.

1.7–6 Make a Venn diagram for four sets.

1.7–7 In a toss of three coins what are the probabilities of 0, 1, 2, or 3 heads? Ans. $1/8, 3/8, 3/8, 1/8$.

1.7–8 What is the probability of drawing either the 10 of diamonds or the jack of spades?

1.7–9 What is the probability of drawing an "ace" (ace= 1), jack, queen or king from a deck of 52 cards? Ans. 4/13.

1.7–10 What is the probability of drawing a black card from the deck?

1.7–11 If two dice are rolled what is the probability both faces are the same? Ans. 1/6.

1.7–12 If three dice are rolled what is the probability that all three faces are the same? Ans. 1/36.

1.7–13 If a coin is tossed n times what is the probability that all the faces are the same? Ans. $1/2^{n-1}$

1.7–14 If one card is drawn at random from each of two decks of cards what is the probability that the two cards are the same? Ans. 1/52.

1.7–15 Odd man out. Three people each toss a coin. What is the probability of some person being the "odd man out"? Ans. 3/4.

1.7–16 If 4 people play odd man out what is the probability that on a round of tosses one person will be eliminated? Ans. 1/2.

1.7–17 On the toss of two dice what is the probabilty that at least one face has a 4 or else that the sum of the faces is 4? Ans. 1/18.

1.7–18 Same as 1.7-17 except the number is 6 (not 4). Ans. 4/9.

1.8 Conditional Probability

Frequently the probability we want to know is *conditional* on some event. The effect of the condition is to remove some of the events in the sample space (list), or equivalently to confine the admissable items of the sample space to a limited region.

Example 1.8–1   *At Least Two Heads in Ten Tosses*

What is the probability of at least two heads in the toss of 10 coins, given that we know that there is at least one head.

The initial sample space of $2^{10} = 1024$ points is uncomfortably large for us to list all the sample points, even when some are excluded; and we do not care to list even the successes. Instead of counting we will calculate the number of successes and the size of the entire sample space.

We may compute this probability in two different ways. First we may reason that the sample space now excludes the case of all tails, which can occur in only one way in the sample space of 2^{10} points; there are now only $2^{10} - 1$ equally likely events in the sample space. Of these there are 10 ways in which exactly one head can occur, namely in any one of the ten possible positions, first, second, ..., tenth. To count the number of events with two or more heads we remove from the counting (but not from the sample space) these 10 cases of a single head. The probability of at least two heads is therefore the ratio of the number of successes (at least 2 heads) to the total number of events (at least one head)

$$(2^{10} - 1 - 10)/(2^{10} - 1) = 1013/1023 = 0.9902\ldots$$

Secondly, we could use a standard method of computing not what is wanted but the opposite. Since this method is so convenient, we need to introduce a suitable notation. If p is the probability of some simple event then we write $q = 1 - p$ as the probability of it not happening, and call it the *complement probability*. Given the above ten coin problem we can compute the complement event, the probability Q that there was one head, and then subtract this from 1 to get

$$P = 1 - Q = 1 - 10/(2^{10} - 1)$$

As a matter of convenient notation we will often use lower case letters for the sample space probabilities and upper case letters for the probabilities of the compound events.

Example 1.8–2   *An Even Sum on Two Dice*

On the roll of two independent dice, if at least one face is known to be an even number, what is the probability of a total of 8?

We reason as follows. One die being even and the sum is to be 8 hence the second die must also be even. Hence the three equally likely successes (the sum is 8) in the sample space are

$$2,6 \qquad 4,4 \qquad 6,2$$

out of a total sample space of $\{36 - (\text{both faces odd})\} = 36 - 3 \times 3 = 27$. Hence the probability is $3/27 = 1/9$.

Example 1.8–3   *Probability of Exactly Three Heads Given that there are at Least Two Heads*

What is the probability of exactly three heads knowing that there are at least two heads in the toss of 4 coins?

We reason as follows. First, how many cases are removed from the original complete sample space $2^4 = 16$ points? There is the single case of all tails, and the four cases of exactly 1 head, a total of 5 cases are to be excluded from the original 16 equally likely possible cases; this leaves 11 still equally likely events in the sample space. Second, for the three heads there are exactly 4 ways that the corresponding 1 tail can arise, ($HHHT$, $HHTH$, $HTHH$, $THHH$), so there are exactly 4 successes. The ratio is therefore $4/11 =$ the probability of exactly three heads given that there are at least two heads.

Exercises 1.8

1.8–1 If a fair coin is tossed 10 times what is the probabilty that the first five are the same side? Ans. $1/16 = 6.25\%$.

1.8–2 Given that the sum of the two faces of a pair of dice is greater than 10, what is the probability that the sum is 12? Ans. $1/3$.

1.8–3 In drawing two cards from a deck (without returning the first card) what is the probability of two aces when you get at least one ace?

1.8–4 What is the probability of different faces turning up on two tosses of a die? Ans. $5/6$.

1.8–5 What is the probability of no two faces being the same on three tosses of a die? Ans. $5/9$.

1.8–6 What is the probability that at least one face is a 6 on the toss of a pair of dice, given that the sum is ≥ 10? Ans. $5/6$.

1.8–7 Given that in four tosses of a coin there are at least two heads, what is the probability that there are at least three heads?

1.8–9 Given that both faces on the toss of two dice are odd, what is the probability of at least one face being a 5? Ans. $5/9$.

1.8–10 In n tosses of a coin, given that there are at least two heads, what is the probability of n or $n-1$ heads?

1.8–11 Given that two cards are honor cards $(10, J, Q, K, A)$ in spades what is the probability that exactly one of them is the ace? Ans. 8/25.

1.8–12 Given that a randomly drawn card is black, what is the probability that it is an honor card? Ans. $10/26 = 5/13$.

1.8–13 Given that an even number of heads show on the toss of 4 coins (0 is an even number), what is the probability of exactly 2 heads? Ans. 3/4.

1.8–14 On the toss of three coins what is the probability of an even number of heads? Ans. 1/2.

1.8–15 On the toss of 4 coins what is the probability of an even number of heads?

1.8–16 From the previous two problems what is the probability of an even number of heads on the toss of n coins? Ans. 1/2.

1.8–19 Given that the sum of the faces on two dice is 6, what is the probability that one face is a 4? Ans. 2/5.

1.8–20 Given that each of the faces on three dice show even number, what is the probability of at least one face being a 6?

1.9   Randomness

Randomness is a negative property; it is the absence of any pattern (we reject the idea that the absence of a pattern is a pattern and for set theory formalists this should be considered carefully). Randomness can never be proved, only the lack of it can be shown. We use the absence of a pattern to assert that the past is of no help in predicting the future, the outcome of the next trial. A random sequence of trials is the realization of the assumption of independence—they are the same thing in different forms. Randomness is "a priori" (before) and not "a posteriori" (after).

We have not defined "pattern." At present it is the intuitive idea that if there is a pattern then it must have a simple description—at least simpler than listing every element of the pattern.

Randomness is a mathematical concept, not a physical one. Mathematically we think of random numbers as coming from a *random source*. Any particular sequence of numbers, once known, is then predictable and hence cannot be random. A reviewer of the famous RAND [R] *Table of a Million Random Numbers* caught himself in mid review with the observation that now that they had been published the numbers were perfectly predictable (if you had the table) and therefore could not be random! Thus the *abstract mathematical concept* of a random sequence of numbers needs to be significantly modified when we actually try to handle particular realizations of randomness.

In practice, of course, you can have only a finite sequence of numbers, and again, of course, the sequence will have a pattern, if only itself! There is a built in contradiction in the words, "I picked a random number between 1 and 10 and got 7." Once selected (a posteriori) the 7 is definite and is not random. Thus the mathematical idea of random (which is a priori) does not match closely what we do in practice where we say that we have a random sample since once obtained the random sample is now definite and is not mathematically random. Before the roll of a die the outcome is any of the 6 possible faces; after the roll it is exactly one of them. Similarly, in quantum mechanics before a measurement on a particle there is the set of possible outcomes (states); after the measurement the particle is (usually) in exactly one state.

On most computers we have simple programs that generate a sequence of "pseudo random numbers" although the word "pseudo" is often omitted. If you do not know that they are being generated by a simple formula then you are apt to think that the numbers are from a random source since they will pass many reasonable tests of randomness; each new pseudo random number appears to be unpredictable from the previous ones and there appears to be no pattern (except that for many random number generators they are all odd integers!), but if you know the kind of generating formula being used, and have a few numbers, then all the rest are perfectly predictable. Thus whether or not a particular stream of numbers is to be regarded as coming from a random source or not depends on your state of knowledge—an unsatisfactory state of affairs!

For practical purposes we are forced to accept the awkward concept of "relatively random" meaning that with regard to the proposed use we can see no reason why they will not perform as if they were random (as the theory usually requires). This is highly subjective and is not very palatable to purists, but it is what statisticians regularly appeal to when they take "a random sample" which once chosen is finite and definite, and is then not random—they hope that any results they use will have approximately the same properties as a complete counting of the whole sample space that occurs in their theory.

There has arisen in computing circles the idea that the measure of randomness (or if you wish, nonrandomness) should be via the shortest program that will generate the set of numbers (without having carefully specified the language used to program the machine!). This tends to agree with many people's intuitive feelings - it should not be easy to describe the lack of a pattern in any simple fashion, it ought to be about as hard as listing all the random numbers. Thus the sequence

$$0 \quad 1 \quad 0 \quad 1 \quad 0 \quad 1 \quad 0 \quad 1 \quad 0 \ldots$$

is "more random" than is the sequence

$$0 \quad 0 \quad 0 \quad 0 \quad 0 \quad 0 \quad 0 \quad 0 \quad 0 \ldots$$

but is less random than the sequence

$$0 \quad 1 \quad 0 \quad 0 \quad 0 \quad 1 \quad 1 \quad 0 \quad 1 \quad 1 \quad 0 \quad 0 \quad 0 \quad 0 \quad 0 \quad 1 \ldots$$

In this approach pseudo random numbers are only slightly random since the program that generates them is usually quite short! It would appear that a genuinely mathematically random number generator cannot be written in any finite number of symbols!

We will use the convenient expression "chosen at random" to mean that the probabilities of the events in the sample space are all the same *unless* some modifying words are near to the words "at random." Usually we will compute the probability of the outcome based on the uniform probability model since that is very common in modeling simple situations. However, a uniform distribution does not imply that it comes from a random source; the numbers $1, 2, 3, \ldots, 6n$ when divided in this order by 6 and the remainders $(0, 1, 2, 3, 4, 5)$ tabulated, gives a uniform distribution but the remainders are not random, they are highly regular!

These various ideas of randomness (likeliness) are not all in agreement with our intuitive ideas, nor with each other. From the sample space approach we see that any hand of 13 cards dealt from a deck of 52 cards is as likely as any other hand (as probable—as random). But the description of a hand as being 13 spades is much shorter than that of the typical hand that is dealt. Furthermore, the shortness of the description must depend on the vocabulary available—of which the game of bridge provides many special words. We will stick to the sample space approach. Note that the probability of getting a particular hand is not connected with the "randomness of the hand" so there is no fundamental conflict between the two ideas.

If this idea of a random choice seems confusing then you can take comfort in this observation: while we will often speak of "picking at random" we will always end up averaging over the whole sample space, or a part of it; we do not actually compute with a single "random sample." This is quite different from what is done in statistics where we usually take a small "sample" from the sample space and *hope* that the results we compute from the sample are close to the ideal of computing over the complete sample space.

Example 1.9–1   *The Two Gold Coins Problem*

There is a box with three drawers, one with two gold coins, one with a gold and a silver coin, and one with two silver coins. A drawer is chosen at random and then a coin in the drawer is chosen at random. The observed coin is gold. What is the probability that the other coin is also gold?

The original sample space *before* the observation is clearly the product space of the three random drawers and the two random choices of which coin, (in order to count carefully we will give the coins marks 1 and 2 when there are two of the same kind, but see Section 2.4)

	Order	
	first/second	first/second
drawer 1	$G_1 G_2$	$G_2 G_1$
drawer 2	GS	SG
drawer 3	$S_1 S_2$	$S_2 S_1$

These six cases, choice of drawer ($p = 1/3$), and then choice of which coin ($p = 1/2$), exhaust the sample space. Thus each compound event has a probability of 1/6. Since, as we observed, high dimensional spaces are hard to draw, we shift to a *listing* of the elementary events as a standard approach.

first drawer	second drawer	third drawer
$G_1 G_2, \ G_2 G_1,$	$GS, \ SG*,$	$S_1 S_2 *, \ S_2 S_1 *$

In this sample space the probability of drawing a gold coin on the first draw (or on the second) is 1/2.

However, the observation that the first drawn coin is gold eliminates the three "starred" possibilities. The remaining three points in the sample space $G_1 G_2, G_2 G_1, GS$, are still equally likely since originally each had the same probability 1/6 so now each has the conditional probability of

Successes/total in the reduced sample space

$$= (1/6)/(1/6 + 1/6 + 1/6) = 1/3$$

Of these three possibilities two give a gold coin on the second draw and only one gives a silver coin, hence the probability of a second gold coin is 2/3.

If this result seems strange to you (after all the second coin is either G or S so why not $p = 1/2$?), then think through how you would do a corresponding experiment. Note the false starts when you get silver coin on the first trial and you have to abandon those trials.

There are many ways of designing this experiment. First imagine 6000 cards, 1000 marked with each of the 6 "initial choice of drawer and coin distribution in the drawer." We imagine putting these 6000 cards in a container, stirring thoroughly, and drawing a card to represent one experimental trial of selecting a drawer and then a coin, and then after looking at the card either discarding the trial if an S showed as the first coin, and if not then going on to see what second coin is marked on the card. We can then return the card, stir the cards and try again and again, until we have done enough trials to convince ourselves of the result. There will, in this version of the experiment, be sampling fluctuations. Here we are (improperly) appealing to you sense that probability is the same as frequency of occurring.

Second, if we go through this mental experiment, but do not return the card to the container, then we will see that after 6000 trials we will have discarded 3000 cards, and of the 3000 we kept 1000 will have the second coin silver, and 2000 will be gold. This agrees with the calculation made above.

A third alternate experiment is to search the container of 6000 cards and remove those for which the silver coin occurs first. Then the remainding 3000 cards will give the right ratio.

Finally there is the very simple experiment, write out one card for each possible situation, remove the failures and simply count, as we did in the above Example 1.9-1.

These thought experiments are one route from the original equally likely sample space to the equally likely censored sample space. The use of mathematical symbols will not replace *your thinking* whether or not you believe that the censoring can affect the *relative* probabilities assigned to the events left; whether or not you believe the above arguments are relevant.

To the beginner the whole matter seems obvious, but the more you carefully think about it the less obvious it becomes (sometimes!). How certain are you that the removal of some cases can not affect the *relative probabilities* of the other cases left in the sample space? In the end it is an assumption that is not mathematically provable but must rest on your intuition of the symmetry and independence in the problem.

Example 1.9–2   *The Two Children Problem*

It is known that the family has two children. You observe one child and that it is a boy, then what is the probability that the other is a boy?

The sample space, listed with the order of observation indicated (first on the left, second on the right), is

$$(B,B) \quad (B,G) \quad (G,B) \quad (G,G)$$

and assuming for the moment both that: (1) the probability of a boy is 1/2 and (2) the sexes in a family are independent, then each point in the sample space occurs with probability $(1/2)(1/2) = 1/4$. The observation that the chosen (first observed) child is a boy eliminates the last two cases, and being equally likely the others both have the conditional probability 1/2. In only one case is the second child a boy, so the probability of the other being a boy is 1/2.

But if you assert only that at least one child in the family is a boy then you remove only one point, GG, from the sample space, and the probability of the other child being a boy is 1/3.

If you are to develop your intuition for probability problems then it is worth your attention to see why the two cases in Example 1.9–2 differ in the result, why in one case the first observation does not affect the second observation while for the conditional probability it does. See also Example 1.9–1. The following Example further illustrates this point.

Example 1.9–3   *The Four Card Deck*

A deck has four cards (either red or black, with face value of 1 or 2). The cards are

$$R1, \quad B1, \quad R2, \quad B2$$

You deal two cards at random.

First question, if one card is known to be a 1, then what is the probability that the other card is also a 1? To be careful we list the complete sample space (of size $4 \times 3 = 12$, the first choice controls the row and the second the column)

$$
\begin{array}{lll}
R1, B1 & R1, R2 & R1, B2 \\
B1, R1 & B1, R2 & B1, B2 \\
R2, R1 & R2, B1 & R2, B2 \\
B2, R1 & B2, B1 & B2, R2
\end{array}
$$

in table form. The fact that there is a 1 (the order of the cards does not matter) eliminates two cases, $R2, B2$ and $B2, R2$, so the sample space is now of size 10. Of these only two cases, $R1, B1$ and $B1, R1$, have a second 1. Hence the probability is $2/10 = 1/5$.

We could also observe at the start that since the order of the cards does not matter then R1,B1 is the same as B1,R1 and that there are then only 6 cases in the sample space

$$R1, B1 \quad R1, R2 \quad R1, B2 \quad B1, R2 \quad B1, B2 \quad R2, B2$$

and each arises by combining two of the original 12 equally likely cases, hence each must now have probability 1/6.

Second question, if the color of the observed card is also known, say, the $R1$, then what is the probability that the other is a 1? In this case only the first row and first column are to be kept, and these total exactly 6 cases.

Of these 6 only 2 cases meet the condition that the second card is a 1. Hence the probability is $2/6 = 1/3$.

The two answers are different, $1/5$ and $1/3$, and the difference is simply that the amount of information (which cases were eliminated from the original sample space) is different in the two examples.

It is important to be sensitive to the effect of different amounts of information in the given statement of the problem so that you develop a feeling for what effects result from slightly different conditions. It is evident that the more restricting the information is, then the more it reduces the sample space and hence can *possibly* change the probability.

Example 1.9–4   *Two Headed Coin*

In a bag of N coins one is known to be a 2-headed coin, and the others are all normal coins. A coin is drawn at random and tossed for k trials. You get all heads. At what k do you decide that it is the 2-headed coin?

To be careful we sketch the sample space

		Trial					
		1st	2nd	3rd	4th	$\cdots$	k^{th}
		H	H	H	H	$\cdots$	H
1 case the 2-headed coin	$p =$	1	1	1	1	$\cdots$	1
$(n-1)$ cases	$p =$	1/2	1/4	1/8	1/16	$\cdots$	$1/2^k$

In particular, for k (heads in a row) we have for the false coin the probability of ($1/n =$ probability of getting the false coin)

$$(1/n)\,(1)$$

while for a good coin

$$\{(n-1)/n\}\,\{1/2^k\}$$

Let us take their ratio

$$\text{false coin}/(\text{good coin}) = 2^k/(n-1)$$

When n is large you need a reasonably large number k of trials of tossing the chosen coin so that you can safely decide that *you probably* have a false coin, (you need $k \sim \log_2(n-1)$ to have the ratio 1), and to have some safety on your side you need more than that number. How many tosses you want to make depends on how much risk you are willing to take and how costly another trial is; there can be no certainty.

Example 1.9–5   *No Information*

You draw a card from a well shuffled deck, but do not look at it. Without replacing it you then draw the second card. What is the probability that the second card is the ace of spades?

The probability that the first card was the ace of spades is $1/52$ and in that case you cannot get it on the second draw. If it was not the ace of spaces, probability $51/52$, then your chance of the ace on the second draw is $1/51$. Hence the total probability is

$$(1/52)(0) + (51/52)(1/51) = 1/52$$

and it is as if the first card had never been drawn! You learned nothing from the first draw so it has no effect on your estimate of the outcome of the second draw.

Evidently, by a slight extension to a deck of n cards and induction on the number of cards removed from the deck and not looked at, then no matter how many cards (less than n) were drawn and not looked at, the probability of then drawing any specified card is $1/n$.

In most situations if one person knows something and uses it to compute a probability then this probability will differ from that computed by another person who either does not have that information or does not use it. See also Example 1.9–8.

Example 1.9–6   *The Birthday Problem*

The famous birthday problem asks the question, "What is the fewest number of people that can be assembled in a room so there is a probability greater than $1/2$ of a duplicate birthday."

We, of course, must make some assumptions about the distribution of birthdays throughout the year. For convenience it is natural to assume that there are exactly 365 days in a year (neglect the leap year effects) and assume that all birthdays are equally likely, namely each date has a probability $1/365$. We also assume that the birthdays are independent (there are no known twins, etc.).

There are the cases of one pair of duplicate birthdays, two pairs of duplicate birthdays, triples, etc.—many different cases to be combined. This is the typical situation where you use the *complement probability* approach and compute the probability that there are no duplicates. We therefore first compute the complement probability $Q(k)$ that k people have no duplicate birthdays.

To find $Q(k)$, the first person can be chosen in 365 equally likely ways—365/365 is the correct probability of no duplication in the selection of only one person. The second person can next be chosen for no duplicate in only 364 ways—with probability $364/365$. The third person must not fall on either of the first two dates so there are only 363 ways, the next 362 ways, ... and the kth in $365 - (k - 1) = 365 - k + 1$ ways. We have, therefore, for these independent selections

$$P(k) = 1 - Q(k)$$

$$= 1 - (365/365)(364/365) \ldots [(365 - k + 1)/365]$$

To check this formula consider the cases: (1), $P(365) \neq 1$; (2), $Q(366) = 0$, hence, as must be, $P(366) = 1$, there is certainly at least one duplicate! Another check is $k = 1$ where $Q(1) = 1$, hence $P(1) = 0$ as it should. Thus the formula seems to be correct.

Alternately, to compute $Q(k)$ we could have argued along counting lines. We count the number of cases where there is no duplicate and divide by the total number of possible cases. The first person can be chosen in 365 ways, the second (non duplicate) in 364 ways, the third in 363 ways, ... the kth in $365 - k + 1$ ways. The total number of ways (the size of the product space) is 365^k and we have the same number for $Q(k)$, the complement probability.

It is difficult for the average person to believe the results of this computation so we append a short table of $P(k)$ at a spacing of 5 and display on the right in more detail the part where $P(k)$ is approximately 1/2.

TABLE 1.9–1

Table of $P(k)$ for selected values

$P(k)$	$\log_{10} P/(1 - P)$	$P(k)$
$P(5) = 0.02714$	-1.55441	$P(21) = 0.44369$
$P(10) = 0.11695$	-0.87799	$P(22) = 0.47570$
$P(15) = 0.25290$	-0.47043	$P(23) = 0.50730$
$P(20) = 0.41144$	-0.15545	$P(24) = 0.53834$
$P(25) = 0.56870$	$+0.12010$	
$P(30) = 0.70632$	0.38113	
$P(35) = 0.81438$	0.64220	
$P(40) = 0.89123$	0.91348	
$P(45) = 0.94098$	1.2026	
$P(50) = 0.97037$	1.5152	
$P(55) = 0.98626$	1.8560	
$P(60) = 0.99412$	2.2281	
$P(65) = 0.99768$	2.6335	
$P(70) = 0.99916$	3.0754	
$P(75) = 0.99972$	3.5527	
$P(80) = 0.99991$	4.0457	
$P(85) = 0.99998$	4.6990	
$P(90) = 0.99999$	4.9999	

    The result that for 23 people (and our assumptions) the probability of a duplicate birthday exceeds $1/2$ is surprising until you remember that any two people can have the same birthday, and it is not just a duplicate of your birthday. There are $C(n, 2) = n(n-1)/2$ pairs of people each pair with a probability of approximately $1/365$ of a coincidence, hence the average number of coincidences is, for $n = 28$,

$$(28 \times 27)/2(365) = 1.0356\ldots$$

hence again the result is reasonable, but see Example 1.9–7.

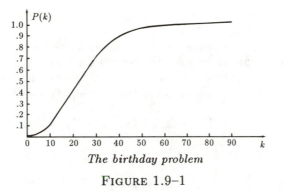

The birthday problem

FIGURE 1.9–1

    The curve of this data is plotted in Figure 1.9–1 and shows a characteristic shape for many probability problems; there is a slow beginning, followed by a steep rise, and then a flattening at the end. This illustrates a kind of "saturation phenomenon"—at some point you pass rapidly from unlikely to very likely. In Figure 1.9–2 we plot $\log\{P/(1-P)\}$ to get a closer look at the two ends of the table.

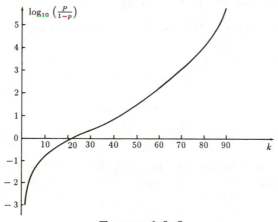

FIGURE 1.9–2

The sequence of numbers we have introduced has the convenient notation of *descending factorials*

$$(n)_k = n(n-1)(n-2) \ldots (n-k+1) \qquad (1.9\text{-}1)$$

which is the product, beginning with n, of k successive terms each 1 less than the preceeding one and ends with $n - (k-1)$.

Example 1.9–7   *The General Case of Coincidences*

If we draw at random from a collection of n distinct items, and replace the drawn item each time before drawing again, what is the probability of no duplicate in k trials?

The reasoning is the same as in the previous problem. The first has a probability of no duplicate is n/n, the second for no duplicate is $(n-1)/n$, and so on to the kth (where you must avoid all the previous $k-1$ samples) which is $(n-k+1)/n$; hence we have the probability for all k independent trials

$$Q_n(k) = n(n-1)(n-2) \ldots (n-k+1)/n^k = (n)_k/n^k$$

$$= n!/n^k(n-k)!$$

$$= [1 - 1/n][1 - 2/n] \ldots [1 - (n-k+1)/n]$$

The numerator in the top of these three equations is the falling (descending) factorial with exactly k terms. For $n = 365$ we have the $Q(k)$ of the birthday problem.

To evaluate this expression easily we use the third line and an inequality from Appendix 1.B, namely that for $x > 0$

$$1 - x < e^{-x}$$

on each term of the above product. This gives

$$Q_n(k) < \exp[-\{1/n + 2/n + \cdots + (k-1)/n\}] = \exp[-k(k-1)/2n]$$

For the probability of the bound to be 1/2 we get

$$ln2 = k(k-1)/2n$$

$$k^2 - k - 2n \; ln \; 2 = 0$$

For the birthday problem $n = 365$, and we solve the quadratic to get 23.000 within roundoff.

To get a feeling for the function $y(k) = (n)_k/n^k$ we look for the place where it rises most steeply. This occurs close to where the second central difference is 0, that is where

$$y(k+1) - 2y(k) + y(k-1) \sim 0$$

We can immediately factor out the $y(k)$ to get (remember $k > 0$)

$$y(k)[(n - k)/n - 2 + n/(n - k + 1)] = 0$$

$$(n - k)^2 + n - k - 2n^2 + 2nk - 2n + n^2 = 0$$

$$k^2 - k - n = 0$$

$$k = \{1 \pm \sqrt{(1 + 4n)}\}/2 = \{1 + \sqrt{(4n + 1)}\}/2$$

$$\sim \sqrt{n} + 1/2$$

For the birthday problem $n = 365$, and we have the approximate place of steepest rise is

$$k \sim \sqrt{n} + 1/2 = 19.10 + 1/2 = 19.6$$

and this indeed we see in Table 1.9–1, and Figure 1.9–1 where the steepest rise precedes the 50% point of $k = 23$.

Example 1.9–8   *The Information Depends on Your State of Knowledge*

There is (was?) a TV game in which you guess behind which of three curtains the prize you might win is placed. Your chance of success is $1/3$ because of the obvious assumption that there is no pattern in the placing of the prize (otherwise long term watchers would recognize it). The next step is that the host of the show pulls back a curtain, other than the one you indicated, and reveals that there is no prize there. You are then given the chance of changing your choice for the payment of a fixed sum. What should you do?

Your action logically depends on what you think the host knows and does. If you assume that the host does not know where the prize is and draws the curtain at random (one of the two left) then the sample space has only six possibilities. If we assume that the prize is behind A but you do not know which is A, (hence any pattern you might use is equivalent to your choosing A, B or C at random), then the sample space (you, host) is

$$(A, B) \quad (A, C) \quad (B, A) \quad (B, C) \quad (C, A) \quad (C, B)$$

each with probability $1/6$. In the case we are supposing, the host's curtain reveals no prize and this eliminates the points (B, A) and (C, A) from the sample space, so that your probability is now

$$P = 2(1/6)/\{1 - 1/3\} = 1/(3 - 1) = 1/2$$

Your probability passed from $1/3$ to $1/2$ because you learned that the prize was not in one of the places and hence it is in one of the remaining two places.

But it is unlikely that the host would ever pull the curtain with the prize behind it, hence your more natural assumption is that he does not pull the curtain at random, but rather knows where the prize is and will never draw that one. Now the situation is that if you happened to pick A (with probability 1/3) the host will pull a curtain at random and the two cases

$$(A, B) \quad \text{and} \quad (A, C)$$

will each have a probability 1/6. But if you pick either B or C then the host is forced to choose the other curtain behind which the prize is not located. Your choices B or C each remain of probability 1/3. Although you now know that the prize is either behind your choice or the other one you have learned nothing since you knew that the curtain drawn would not show the prize. You have no reason to change from you original choice having probability of 1/3. Hence the other curtain must have probability 2/3.

If this seems strange, let us analyse the matter when the host chooses with a probability p the curtain that does not have the prize. Then the host chooses the curtain with the prize $q = 1 - p$. Now the cases in the sample space are:

$$p(AB) = 1/6 = p(AC)$$

$$p(BA) = (1 - p)/3 = p(CA)$$

$$p(BC) = p/3 = p(CB)$$

As a check we see that the total probability is still 1. The fact that the curtain was drawn and did not reveal the prize means that the cases $p(BA)$ and $p(CA)$ are removed from the sample space. So now your probability of winning is

$$P = \{p(AB) + p(AC)\}/\{1 - p(BA) - p(CA)\} = \{1/3\}/\{1 - 2(1 - p)/3\}$$

$$= \{1/3\}/\{1/3 + 2p/3\} = 1/\{1 + 2p\}$$

We make a short table to illustrate things and check our understanding of the situation.

p	P	meaning
1	1/3	Host never reveals the prize.
1/2	1/2	Host randomly reveals the prize.
0	1	Host must reveal it if it can be done and since he did not you must have won.

Hence what you think the host does influences whether you should switch your choice or not. The probabilities of the problem depend on your state of knowledge.

It should be evident from the above Examples that we need to develop systematic methods for computing such things. The patient reasoning we are using will not work very well in larger, more difficult problems.

Exercises 1.9

1.9–1 Consider the birthday problem except that you ask for duplicate days of the month (assume that each month has exactly 30 days). Ans. $P(7) = 0.5308$

1.9–2 If there are only 10 possible equally likely outcomes how long will you expect to wait until the probability of a duplicate is $> 1/2$? Assume that the trials are independent. Make the complete table. Ans. $P(1) = 0, P(2) = .1, P(3) = .28, P(4) = .496, P(5) = .6976, P(6) = .8488, P(7) = .93952, P(8) = .981856, P(9) = .9963712, P(10) = .99963712$.

1.9–3 There are three children in a family and you observe that a randomly chosen one is a boy, what is the probability that the other two children have a common sex?

1.9–4 There is a box with four drawers. The contents are respectively GGG, GGS, GSS, and SSS. You pick a drawer at random and a coin at random. The coin is gold. What is the probability that there is another gold coin in the drawer? That a second drawing from the drawer will give a gold coin? Ans. 5/6, 2/3.

1.9–5 Generalize the previous problem to the case of n coins per drawer. Find the probability that the first two coins drawn are both gold. Ans. 2/3

1.9–6 If you suppose that a leap year occurs exactly every fourth year, what is the probability of a person being born on Feb. 29?

1.9–7 In Example 1.9–6 use $y'' = 0 (y'' = $ second derivative$)$ in place of the second difference to obtain a similar result.

1.9–8 You have $2n$ pieces of string hanging down in your hand and you knot pairs of them, tying at random an end above with and end above and below with below, also at random. What is the probability that you end up with a single loop? [*Hint:* Knotting pairs at random above has no effect on the problem and you are reduced to n U shaped pieces of string. Proceed as in the birthday problem with suitable modifications (of course)].

1.9–9 From a standard deck k cards are drawn at random. What is the probability that no two have the same number? Ans. $4^k (13)_k /(52)_k$

1.9–10 One card is drawn from each of two shuffled decks of cards. What is the probability:
  1. the card from the first deck is black?
  2. at least one of the two cards is black?
  3. if you know that at least one is black, that both are black?
  4. that the two cards are of opposite color?

1.9–11 In Appendix 1.B get one more term in the approximation for $1 - x$ using $e^{-x} \exp(-x^2/2)$.

1.9–12 Apply Exercise 1.9–11 to Example 1.9–7.

1.9–13 There are w white balls and b black balls in an urn. If $w + b - 1$ balls are drawn at random and not looked at what is the probability that the last ball is white? Ans. $w/(w + b)$.

1.9–14 Show that a bridge hand can be easily described by 64 bits. [*Hint:* give three 4 bit numbers to tell the number of cards in spades, hearts, and diamonds (the number of clubs is obvious then) and then list the 13 card face values in the suits. The minimum representation is much harder to convert to and from the hand values.]

1.10 Critique of the model

Let us review this model of probability. It uses only the simple concepts of symmetry and interchangeability to assign a probability measure to the finite number of possible events. The concept of the sample space is very useful, whether we take the entire sample space, which may often be built up by constructing the product space from the simpler independent sample spaces, or take some subspace of the sample space.

The model of equally likely situations, which is typical of gambling, is well verified in practice. But we see that all probabilities must turn out to be rational numbers. This greatly limits the possible applications since, for example, the toss of a thumb tack to see if the point is up or it is on its side (Figure 1.10–1) seems unlikely to have a rational number as its probability. Furthermore, there are many situations in which there are a potentially infinite number of trials, for example tossing a coin until the first head appears, and so far we have limited the model to finite sample spaces. It is therefore clear that we must extend this model if we are to apply probability to many practical situations; this we will do in subsequent chapters beginning with Chapter 4.

Thumb tack

FIGURE 1.10–1

It is difficult to quarrel with this model on its own grounds, but it is still necessary to connect this model with our intuitive concept of probability as a long term frequency of occurrence, and this we will do in the next chapter where we develop some of the consequences of this model that involve more mathematical tools. The purpose of separating the concepts of probability from the mathematics connected with it, is both for philosophical clarity (which is often sadly lacking in many presentations of probability) and the fact that the mathematical tools have much wider applications to later models of probability that we will develop.

Remember, so far we have introduced a formal measure of probability based on symmetry and it has no other significant interpretation at this point. We used this measure of probability to show how to compute the probabilites of more complex situations from the uniform probability of the sample space. As yet we have shown no relationship to the common view that probability is connected with the long term ratio of successes to the total number of trials. Although very likely you have been interpreting many of the results in terms of "frequencies," probability is still a measure derived from the symmetry of the initial situation. The frequency relationship will be derived in Section 2.8.

Appendix 1.A   *Bounds on Sums*

We often need to get reasonable bounds on sums that arise in probability problems. The following is a very elementary way of getting such bounds, and they are often adequate.

Suppose we have the continuous function

$$y = f(x)$$

and want to estimate the sum

$$S(N) = \sum_{n=1}^{N} f(n)$$

We also suppose that the second derivative $f''(x)$ is of constant sign.

Suppose first that

$$f''(x) > 0$$

The trapezoid rule overestimates the integral, Figure 1.A–1,

$$\int_1^N f(x)\,dx \le \frac{1}{2}f(1) + f(2) + \cdots + \frac{1}{2}f(N)$$

Hence add $(1/2)[f(1) + f(0)]$ to both sides to get

$$\int_1^N f(x)\,dx + \{f(1) + f(N)\}/2 \le \sum_1^N f(n) \qquad (1.A\text{--}1)$$

On the other hand the midpoint integration formula underestimates the integral. We handle the two half intervals at the ends by fitting the tangent

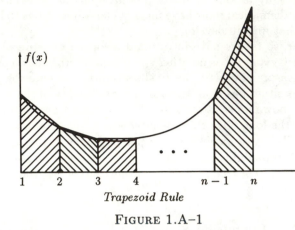

Trapezoid Rule

FIGURE 1.A–1

line at the ends (see Figure 1.A–2).

$$\int_1^N f(x)\,dx \geq \int_1^{3/2} \{f(1) + (x-1)f'(1)\}\,dx$$

$$+ f(2) + f(3) + \cdots + f(N-1) + \int_{N-1/2}^{N} \{f(N) + (x-N)f'(N)\}\,dx$$

$$\geq f(1)/2 + (1/2)^2 f'(1)/2 + \sum_{2}^{N-1} f(n) + f(N)/2 - (1/2)^2 f'(N)/2$$

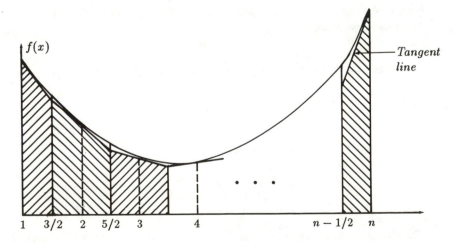

FIGURE 1.A–2

Rearranging things we have

$$\sum_{1}^{N} f(n) \le \int_{1}^{N} f(x)\,dx + [f(1) + f(N)]/2 + [f'(N) - f'(1)]/8 \qquad (1.A-2)$$

If $f''(x) < 0$ the inequalities are reversed.

The two bounds 1.A–1 and 1.A–2 differ by the term

$$[f'(N) - f'(1)]/8 \qquad (1.A-3)$$

Since most of the error generally occurs for the early terms you can sum the first k terms separately and then apply the formulas to the rest of the sum. Often the result is much tighter bounds since $f'(1)$ becomes $f'(k+1)$ in the error term and $f'(k)$ is generally a decreasing function of k.

We now apply these formulas to three examples which are useful in practice. First we choose $f(x) = 1/x$. We have $f'(x) = -1/x^2$ and $f''(x) = 2/x^3 > 0$. The indefinite integral is, of course, merely $ln\,N$. Hence from 1.A–1 and 1.A–2 we have for the *harmonic series*

$$ln\,N + (N+1)/2N \le \sum_{1}^{N} 1/n \le ln\,N + (N+1)/2N + (1 - 1/N^2)/8 \quad (1.A-4)$$

The sum

$$\sum_{n=1}^{N} 1/n = H(N) \qquad (1.A-5)$$

occurs frequently, hence a short table of the exact values is useful to have. Similarly, the sums

$$\sum_{n=1}^{N} 1/n^2 = D(N) \qquad (1.A-6)$$

are also useful to have. Since most of the error arises from the early terms, using these exact values and then starting the approximation formulas at the value $n = 11$ will give much better bounds.

A Short Table of H(N) and D(N)

N	H(N)		D(N)
	fraction	decimal	decimal
1	1	1.00000	1.00000
2	3/2	1.50000	1.25000
3	11/6	1.83333	1.36111
4	25/12	2.08333	1.42361
5	137/60	2.28333	1.46361
6	147/60	2.45000	1.49133
7	1089/420	2.59285	1.51180
8	2283/840	2.71786	1.52742
9	7129/2520	2.82897	1.53977
10	7391/2520	2.92897	1.54977

At $N = 10$ the limits of the bounds are $2.85207 < H(N) < 2.97634$. The limiting value of $D(N) = \pi^2/6 = 1.64493\ldots$.

There is also a useful analytic expression for $H(N)$ which we will not derive here, namely

$$H(N) = \ln N + \gamma + 1/2N - 1/12N^2 + 1/120N^3 + \cdots \qquad (1.A\text{--}7)$$

where $\gamma = 0.57721\,56649\ldots$ is Euler's constant. If we use this formula through the $1/N^2$ term for $N = 10$ we get $H(10) = 2.92897$ which is the correct rounded off number.

For the second example we use $f(x) = 1/x^2$, for which $f'(x) = -2/x^3$ and $f''(x) = 6/x^4 > 0$. The formulas give

$$3/2 - 1/N + 1/2N^2 \leq \sum_{1}^{N} 1/n^2 \leq 3/2 - 1/N + 1/2N^2 + (N^3 - 1)/N^3 \quad (1.A\text{--}8)$$

For the third example we pick $f(x) = \ln x$, for which $f'(x) = 1/x$ and $f''(x) = -1/x^2 < 0$. Hence the inequalities are reversed. We get for the integral (using integration by parts)

$$\int_{1}^{N} \ln x \, dx = N \ln N - N + 1$$

and for the formula for the bounds

$$N \ln N - N + 1 + (1/2)\ln N \geq \sum_{n=1}^{N} \ln n = \ln N!$$

$$\geq N \ln N - N + (1/2)\ln N + 7/8 + 1/8N$$

Dropping the last term on the right strengthens the inequality and taking exponentials we get the more usual form for the factorial function

$$N^{N+1/2}e^{-N}e \geq N! \geq N^{N+1/2}e^{-N}e^{7/8} \qquad (1.\text{A}-9)$$

These bounds may be compared with the asymptotic form of Stirling's factorial approximation

$$N! = N^N e^{-N}\sqrt{(2\pi N)} \qquad (1.\text{A}-10)$$

Note that $e = 2.71828$, $\sqrt{(2\pi)} = 2.50663$, and $e^{7/8} = 2.39887$ (all to five decimal places).

Bounds are almost essential in "deep" computations where approximations are combined in many ways. On the other hand for "shallow" computations Stirling's and other approximations are often very useful.

A particularly simple approximation is the midpoint formula

$$\int_{a-1/2}^{b+1/2} f(x)\, dx \sim \sum_{n=a}^{n=b} f(n) \qquad (1.\text{A}-11)$$

In the three earlier cases we get for $f(x) = 1/x$

$$\sum_{1}^{N} 1/n \sim ln(2N+1) \qquad (1.\text{A}-12)$$

for $f(x) = 1/x^2$ we get

$$\sum_{1}^{N} 1/n^2 \sim 4N/(2N+1) \qquad (1.\text{A}-13)$$

and for $f(x) = ln\, x$ we get

$$\sum_{1}^{N} ln\, n \sim (N+1/2)ln(N+1/2) - N + (1/2)ln\, 2 \qquad (1.\text{A}-14)$$

Taking exponentials we get the formula

$$n! \sim (N+1/2)^{N+1/2}e^{-N}\sqrt{2} \qquad (1.\text{A}-15)$$

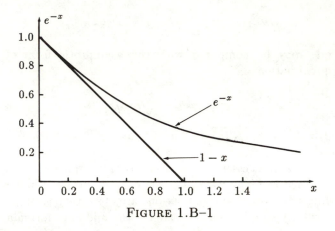

$$\text{Figure 1.B--1}$$

Appendix 1.B   *A Useful Bound*

If we compute the tangent line to the curve

$$y(x) = e^{-x} = \exp(-x)$$

at $x = 0$ we get for the first derivative

$$y'(x) = e^{-x}$$

hence

$$y'(0) = -1$$

and the tangent line is

$$y(x) - 1 = (-1)(x - 0)$$

$$y = 1 - x.$$

Since $y''(x) = \exp(-x) > 0$ we deduce, or else from a sketch of the curve Figure 1.B–1,

$$e^{-x} = \exp(-x) \geq 1 - x \tag{1.B--1}$$

2

Some Mathematical Tools

2.1 Introduction

C. S. Peirce (1839–1914) observed [N, p.1334] that:

> *"This branch of mathematics [probability] is the only one, I believe, in which good writers frequently get results entirely erroneous. In elementary geometry the reasoning is frequently fallacious, but erroneous conclusions are avoided; but it may be doubted if there is a single extensive treatise on probabilities in existence which does not contain solutions absolutely indefensible. This is partly owing to the want of any regular methods of procedure; for the subject involves too many subtleties to make it easy to put problems into equations without such aid."*

There were, I believe, two additional important reasons for this state of affairs at that time and to some extent even now; first there was a lack of clarity on just what model of probability was being assumed and of its known faults, and second, there was a lack of intuition to protect the person from foolish results. Feller [F, p.67] is quite eloquent on this point. Thus among the aims of this book are to provide: (1) careful discussions of the models assumed and any approach adopted; (2) the use of "regular methods" in preference to trick methods that apply to isolated problems; (3) the deliberate development of intuition by the selection of problems, (4) the analysis of the results, and (5) the systematic use of reasonableness tests of equations and results. By these methods we hope to mitigate the statement of C. S. Peirce just quoted.

In the previous chapter we created a probability model based on symmetry, and noted that it could only give rational values for the assignment of the probability to the events, and that all the elementary events had the same probability. We will later be able to handle a wider range of probabilities, namely any real number between 0 and 1 (and complex numbers in Section 8.13) as well as nonuniform probability distributions, hence we will now assume this use and not have to repeat the development of the mathematical tools for handling probability problems. The relevant rules we have developed apply to these nonrational probabilities as can be easily seen by rereading the material.

We saw that the central problem in computing the probability of a complex event is to find all the equally likely elementary successful events in the sample space (those that meet the given conditions), or else the failing events (those that do not), count the successes, or failures, and then divide by the size of the whole sample space. Instead of counting the successes in the uniform sample space and dividing by the total we can add the probabilities of all the individual successes; this is equivalent to assigning to each point in the uniform sample space the probabilty 1/(total number of points in the sample space).

If the probabilities are not uniform in the sample space we cannot simply count and divide by the total, but as just noted we must add the probabilities of the individual successful events to get the total probability to assign to the complex event since the probability of the sum is additive (we pick our sample space points to be independent hence the probability of a complex event involving sums of points is the sum of the probabilities of its independent parts). Since the total probability is 1 there is no need to divide by the sum of the probabilities (which is 1).

We also saw that the probabilities assigned to the points in the product space could often be found as the product of the probabilities of the basic events that make up the corresponding point in the product space. We adopt a colorful language to use while computing probability problems; we say, for example, "Select at random..." and mean "Count the number of successes (in the case of a uniform probability assignment)." As an example, "Select a random day in a year of 365 days." means that since any day of the year meets this condition there are 365 choices, and the corresponding probability that you select a (some) day is $365/365 = 1$. The probability of getting a specific day *if it is named in advance* is, of course, 1/365. If we want to select at random, from the possible 365 days in a year, a day that falls in a certain month of 30 days, then there are only 30 successful possible selections, and the corresponding probability is $30/365 = 6/73$. We say that we can select at random a day in the year that falls in the given month in 30 successful ways, hence with probability of 30/365.

The language "select at random" enables us to pass easily from the uniform probability spaces to nonuniform probability spaces, since in both cases we use the same colorful language and the results are the same; of course for the nonuniform case we must add the probabilities of the successes.

A "random selection" is only a colorful way of talking, and when the word "random" is used without any modifier it implies that the probability assignment is uniform.

     The purpose of this chapter is to introduce the main tools for computing simple probability problems which have a finite, discrete sample space, and to begin the development of uniform methods of solution as well as the development of your intuition. This Chapter also shows a connection to the frequency approach to probability. Chapter 3 will further develop, in a more systematic way, the mathematical tools needed for finite sample spaces. There is a deliberate separation between the model of probability being assumed, the concepts needed for solving problems in that area, and the mathematical tools needed for their solution.

2.2 Permutations

You have already seen that the main difficulty, in simple probability problems, is to count the number of ways something can be done. Hence we begin with this topic (actually partially repeat).

     Suppose you have a collection of n distinct things (unspecified in detail, but whatever you care to think about). From these n items you then select at random k times, replacing the selected item each time before the next selection. The sample space is the product space $n \times n \times n \ldots \times n$ (k times) with exactly (we "star" important equations)

$$n^k \tag{2.2-1}*$$

items in the product space. This is called *sampling with replacement*. The probability distribution for a *random selection* is uniform, hence each item in the sample space will have the probability $1/n^k$.

     Again, suppose you have n things and sample k times, but now you do not replace the sampled items before the next selection. When the *order of selection is important* then we call the selection without replacement a *permutation*. The number of these permutations is written as

$$P(n, k)$$

and is called "the permutation of n things taken k at a time." To find the numerical value for a random (uniform) selection we argue as before, (Examples 1.9–6 and 1.9–7); the first item may be selected in n ways, the second distinct item in $n - 1$ ways, the third in $n - 2$ ways, and the kth in $n - k + 1$ ways. The product space is of size

$$P(n, k) = n(n - 1)(n - 2) \ldots (n - k + 1) = (n)_k \tag{2.2-2}*$$

where $(n)_k$ is the standard notation for the *falling factorial*, (see Equation 1.9–1). By our method of selection of the individual cases in the $(n)_k$ are uniformly probable hence we add the number of cases to get the $P(n, k)$.

To partially check this formula we note that when $k = 1$ the answer is correct, and when $k = n + 1$ we must have zero since it is impossible to select n + 1 items from the set of n items.

A useful bound on $P(n, k)$ can be found by multiplying and dividing the right hand side of (2.2–2) by n^k

$$P(n, k) = n^k(1)(1 - 1/n)(1 - 2/n)\ldots(1 - (k - 1)/n)$$

and then using the result in Appendix 1.B (see also Example 1.9–7)

$$1 - x \le e^{-x}$$

for each factor $(1 - i/n)$, $(i = 1, 2, \ldots, k - 1)$. Summing this arithmetic progression in the exponent

$$-\{1/n + 2/n + \cdots + (k - 1)/n\} = -k(k - 1)/2n$$

we get the result

$$P(n, k) \le n^k \exp\{-k(k - 1)/2n\} \tag{2.2–3}$$

For an alternate proof that all the $P(n, k)$ individual cases are uniformly probable note that from the sample space of all possible choices of k items, which we saw (2.2–1) was uniformly probable and of size n^k, we excluded all those which have the same item twice or more, and we deduced that there are exactly $P(n, k)$ such points left in the sample space. Since a random choice leads to a uniform distribution in the original product space, the distribution is still uniform after the removal of the samples with duplicates (see Example 1.9–1, the gold coin problem). All the permutations in the $P(n, k)$ are equally likely to occur when the items are selected at random. The exponential gives an estimate of the modifying factor for the number of terms in going from the sample space of sampling with replacement to sampling without replacement.

The expression $P(n, k)$ is the beginning of a factorial. If we multiply both numerator and denominator by $(n - k)!$ the numerator becomes $n!$, and we have

$$P(n, k) = \frac{n!}{(n - k)!} \tag{2.2–4}*$$

as a useful formula for the permutations of n things taken k at a time. Recall that by convention $0! = 1$ and that $P(n, n) = n!$. We need also to observe that when $k = 0$ we have

$$P(n, 0) = \frac{n!}{n!} = 1 = (n)_0 \tag{2.2–5}$$

At first this seems like a curious result of no interest in practice, but when programming a descending factorial on a computer you *initialize* the iterative loop of computation for P(n,k) with P(n,0) = 1; then and only then will each cycle of the computing loop give the proper number and be suitably recursive.

Example 2.2–1   *Arrange Three Books on a Shelf*

In how many ways can you arrange 3 books on a shelf out of 10 (distinct) books? Since we are assuming that the order of the books on the shelf matters (the word "arrange"), we have

$$P(10,3) = 10 \times 9 \times 8 = 720$$

ways.

Example 2.2–2   *Arrange Six Books on a Shelf*

In how many ways can you arrange on a shelf 6 books out of 10 (distinct) books? You have

$$P(10,6) = 10 \times 9 \times 8 \times 7 \times 6 \times 5 = 151,200$$

possible ways of arranging them.

Example 2.2–3   *Arange Ten Books on a Shelf*

In how may ways can you arrange on a shelf a set of 10 distinct books? You have

$$P(10,10) = 10! = 3,628,800$$

and you see how fast permutations can rise as the number of items selected increases. The Stirling approximation from Appendix 1.A gives $10! \sim 3,598,695.6$ and the ratio of Stirling to true is 0.99170.

Example 2.2–4   *Two Out of Three Items Identical*

Suppose you have three items, two of which are indistinguishable. How many permutations can you make? First we carefully write out the sample space with elements a_1, a_2, b supposing we have three distinct elements.

$$a_1 a_2 b$$
$$a_1 b a_2$$
$$a_2 a_1 b$$
$$a_2 b a_1$$
$$b a_1 a_2$$
$$b a_2 a_1$$

Now if a_1 and a_2 are indistiguishable then items on lines 1 and 3, 2 and 4, and 5 and 6 are each the same; the reduction of the sample space from 6 to 3 is *uniform*. The following formula gives the proper result

$$3!/2 = 6/2 = 3$$

Example 2.2–5   *Another Case of Identical Items*

In permutation problems there are often (as in the previous Example) some indistinguishable items. For example you may have 7 items, a, a, a, b, b, c, d. How many permutations of these 7 items are there?

We again attack the problem by throwing it back on known methods; we first make the 3 a's distinct, calling them a_1, a_2, a_3, and similarly the 2 b's are now to be thought of as b_1, and b_2. Now we have $P(7,7) = 7!$ permutations. But of these there are $P(3,3) = 3!$ permutations which will all become the same when we remove the distinguishing subscripts on the a's, and this applies *uniformly* throughout the sample space, hence we have to divide the number in the total sample space by $P(3,3) = 3!$, Similarly, for the b's we get $P(2,2) = 2!$ as the dividing factor. Hence we have, finally, the sample space of distinct permutations

$$\frac{7!}{3!2!1!1!} = 7 \times 6 \times 5 \times 4/2 = 420$$

as the number of permutations with the given repetitions of indistinguishable items (we have put 1! twice in the denominator for symmetry and checking reasons, $3 + 2 + 1 + 1 = 7$). Since the reduction from distinct items in the original permutation sample space to those in the reduced permutation space is *uniform* over the whole sample space, the probabilities of the random selection of a permutation with repetitions are still uniform. If this is not clear see the previous Example 2.2–4.

The general case is easily seen to follow from these two special cases, Examples 2.2–4 and 2.2–5. If we have n_1 of the first kind, n_2 of the second, ..., and n_k of the kth, and if

$$n_1 + n_2 + \cdots + n_k = n$$

then there are

$$\frac{n!}{n_1!n_2!\ldots n_k!} = C(n; n_1, n_2, \ldots n_k) \qquad (2.2\text{–}6)^*$$

permutations all of equal probability of a random selection.

These numbers, $C(n; n_1, n_2, \ldots n_k)$, are called the *multinomial coefficients* in the expansion

$$(t_1 + t_2 + t_3 + \cdots + t_k)^n$$

They occur when you expand the multinomial and select the term in t_1 to the power n_1, t_2 to the power $n_2, \ldots, t_k$ to the power n_k, and then examine its coefficient. The coefficient is the number of ways that this combination of powers can arise; this coefficient is the number of permutations with the given duplication of items. Note that if $n_1 = k$ and all the other $n_i = 1$, then this is $P(n, k)$.

This formula applies to the case when all the terms are selected. If only a part of them are to be selected then it is much more complex in detail, but the ideas are not more complex. You have to eliminate the "over counting" that occurs when the items are first thought of being distinct because the reductions are not necessarily uniform over the sample space. We will not discuss this further as it seems to seldom arise in practice. Simple cases can be done by merely listing the sample space.

Exercises 2.2

2.2-1 How many arrangements on a platform can be made when there are 10 people to be seated?

2.2-2 How many arrangements can you make by selecting 3 items from a set of 20? Ans. 570.

2.2-3   In a standard deck of 52 cards what is the probability that on drawing two cards you will get the same number (but not suit)? Ans. 3/51.

2.2-4 In drawing three cards what is the probability that all three are in the same suit? Ans. $(12/51)(11/50) = 22/425$.

2.2-5 In drawing $k \leq 13$ cards what is the probability that all are in the same suit?

2.2-6 Show that $P(n, k) = nP(n - 1, k - 1)$.

2.2-7 You are to place 5 distinct items in 10 slots; in how many ways can this be done?

2.2-8 Tabulate $P(10, k)/10^k$ for $k = 0, 1, \ldots 10$. Ans. 1, 0.9, 0.72, 0.504, etc.

2.2-9 Estimate $P(100, 10)$ using (2.2-3).

2.2-10 Estimate $P(100, 100)$ using both (2.2-3) and Stirling's formula (1.A-10). Explain the difference.

2.2-11 Given a circle with n places on the circumference, in how many ways can you arrange k items? Ans. $P(n, k)/n$

2.2-12 If in 2.2-11 you also ignore orientation, and n is an odd number, in how many ways can you arrange the k items?

2.2-13 Using all the letters how many distinct sequences can you make from the letters of the word "success"? Ans. 420.

2.2-14 Using all the letters how many distinct sequences can you make from the letters of the word "Mississippi"? Ans. $C(11; 4, 4, 2, 1) = 4950$.

2.2-15 Using all the letters how many distinct sequences can you make from the letters of the word "Constantinople"?

2.2-16 How many distinct three letter combinations can you make from the letters of Mississippi? Ans. 38.

2.2-17 How many four letter words can you make from the letters of Mississippi?

2.2-18 If an urn has 7 w (white) and 4 b (black) balls show that on drawing two balls $Pr\{w, w\} = 21/55$, $Pr\{b, w\} = 14/55 = Pr\{w, b\}$, $Pr\{b, b\} = 6/55$.

2.2–19   If you put 5 balls in three bins at random what is the probablity of exactly 1 empty bin?

2.2–20   There are three plumbers in town and on one day 6 people called at random for a plumber. What is the probability 3, 2 or only one plumber was called? Show that the $Pr\{3, 2, 1\} = 20/343$.

2.2–21   Discuss the accuracy of (2.2–3) when k is small with respect to n. When k is large.

2.3   Combinations

A *combination* is a permutation when *the order is ignored*. A special case of the multinomial coefficients occurs when there are only two kinds of items to be selected; they are then called *the binomial* (bi = two, nomial = term) *coefficients* $C(n, k)$. The formula (2.2–6) becomes

$$C(n, k) = \frac{n!}{k!(n - k)!} = C(n, n - k) \qquad (2.3\text{–}1)^*$$

where there are k items of the first kind (selected) and $n - k$ items of the other kind (rejected). We use the older notation C(n,k) in place of the currently popular notation

$$\binom{n}{k}$$

because: (1) it is easier to type (especially for computer terminals), (2) does not involve a lot of special line spacing when the symbol occurs in the middle of a line of type, and (3) simple computer input and output programs can recognize and handle it easily when it occurs in a formula. Note that (2.3–1) is a special case of (2.2–6) when you use $k_2 = n - k_1$. Note also (2.3–1) says that what you select is uniquely determined by what you leave.

Recurrence relations are easy to find for the $C(n, k)$ and often shed more light on the numbers than an actual table of them. For the index k we have, by adjusting the factorials,

$$C(n, k + 1) = \frac{n!}{(k + 1)!(n - k - 1)!}$$

$$= \frac{n!(n - k)}{(k + 1)k!(n - k)(n - k - 1)!} \qquad (2.3\text{–}2)^*$$

$$= \frac{(n - k)}{(k + 1)} C(n, k)$$

Clearly the largest $C(n, k)$ occurs around $(n - k)/(k + 1) \sim 1$, that is $k \sim (n - 1)/2$.

For the index n we have the recurrence relation

$$C(n + 1, k) = \frac{(n + 1)!}{k!(n + 1 - k)!} = \frac{(n + 1)n!}{k!(n + 1 - k)(n - k)!}$$

$$= \frac{(n + 1)}{(n + 1 - k)} C(n, k)$$

$(2.3\text{-}3)^*$

The binomial coefficients $C(n, k)$ arise from

$$(t_1 + t_2)^n = \sum_{k=0}^{n} C(n, k) t_1^{n-k} t_2^k$$

In a more familiar form we have $(t_1 = 1, t_2 = t)$

$$(1 + t)^n = 1 + nt + \frac{n(n - 1)}{2} t^2 + \cdots + t^n$$

$$= \sum_{k=0}^{n} C(n, k) t^k$$

$(2.3\text{-}4)^*$

where the $C(n, k)$ are the number of ways k items can be selected out of n without regard to order. The equation $(2.3\text{-}4)$ *generates* the *binomial coefficients*.

From this generating function $(2.3\text{-}4)$ and the observation that

$$(1 + t)^{n+1} = (1 + t)(1 + t)^n$$

we can equate like powers of t^k on both sides (since the powers of t are linearly independent) to get the important relation

$$C(n + 1, k) = C(n, k) + C(n, k - 1) \qquad (2.3\text{-}5)^*$$

with the side conditions that

$$C(n, 0) = C(n, n) = 1$$

An alternate derivation goes as follows. Suppose we have $n + 1$ items. All possible subsets of size k can be broken into two classes, those without the $(n + 1)$st item and those with it. Those without the $(n + 1)$st item total simply $C(n, k)$, while those with it require selection only $k - 1$ more items, and these total $C(n, k - 1)$. Thus we have $(2.3\text{-}5)$.

This identity leads to the famous Pascal triangle where each number is the sum of the two numbers immediately above, and the edge values are all 1.

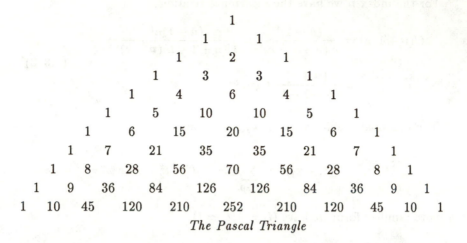

The Pascal Triangle

It might be thought that to get a single line of the binomial coefficients of order n the triangle computation would be inefficient as compared to the recurrence relation (2.3–2). If we estimate the time of a fixed point multiplication as about 3 additions and a division as about two multiplications, then we see that for each term on the nth line we have two additions, one multiplication and one division, or about 11 addition times per term. There being $n - 1$ terms to compute on the nth line we have to compare this with the triangle down to the nth line, namely $n(n - 1)/2$ additions. This leads to comparing

$$n/2 \quad \text{with} \quad 11 \longrightarrow n = 22$$

(Floating point arithmetic would give, of course, different results.) This suggests that as far as the amount of computing arithmetic is concerned it is favorable to compute the whole triangle rather than the one line you want until $n = 22$—which is quite surprising and depends, of course, on the actual machine times for the various operations. Symmetry reduces the amount of computation necessary by 1/2 in both approaches. One can squeeze out time for the one line approach by various tricks depending on the particular machine—the point is only that the Pascal triangle is surprisingly efficient on computers as contrasted with human computation.

From the *generating function* (2.3–4) we can get a number of interesting relationships among the binomial coefficients.

Example 2.3–1   *Sums of Binomial Coefficients*

If we set $t = 1$ in (2.3–2) we get

$$(1+1)^n = \sum_{k=0}^{n} C(n,k) = 2^n \qquad (2.3\text{–}6)^*$$

In words, the sum of all the binomial coefficients of index n is exactly 2^n.

If we set $t = -1$ we get the corresponding sum with alternating signs

$$\sum_{k=0}^{n} (-1)^k C(n,k) = 0 \qquad (2.3\text{–}7)$$

The alternating sum of the binomial coefficients is exactly 0 for all n. Thus the sum of all the even indexed coefficients is the same as the sum of all the odd indexed coefficients.

If we differentiate the generating function (2.3–4) with respect to t we get the identity

$$n(1+t)^{n-1} = \sum_{k=1}^{n} k C(n,k) t^{k-1}$$

and when we put $t = 1$ we get

$$n 2^{n-1} = \sum_{k=1}^{n} k C(n,k) \qquad (2.3\text{–}8)^*$$

Many other useful relationships can be found by: (1) suitably picking a function of t to multiply through by, (2) differentiating or integrating one or more times, and finally (3) picking a suitable value for t. The difficulty is to decide what to do to get the identity you want. See Appendix 2.B.

Example 2.3–2   *Bridge Hands*

In the game of bridge each hand is dealt 13 cards at random from a deck of 52 cards with four suits. How many different sets of 4 hands are there (the order of the receiving the cards does not matter)?

Evidently we have the combination (since order in the hands does not matter)

$$52!/(13!)^4 = 5.36447\ldots \times 10^{28}$$

when you assume that the hands are given but not the positions around the bridge table. Thus you see the enormous number of possible bridge dealings. The Stirling approximation (A.1) gives $5.49\ldots \times 10^{28}$.

Example 2.3–3   *3 Out of 7 Books*

A student grabs at random 3 of his 7 text books and dashes for school. If indeed the student has 3 classes that day what is the probability that the correct three books were selected?

We can argue in either of two ways. First, we can say that there are exactly $C(7,3)$ equally likely selections possible out of the set of 7 books and that only 1 combination is correct, hence the probability is

$$1/C(7,3) = 3!/(7 \times 6 \times 5) = 1/35$$

We can also argue (repeating the basic derivation of the binomial coefficients) that the first book can be successfully selected in 3 out of 7 ways, hence with probability 3/7. Then the second book can independently be successfully selected in 2 out of 6 ways with probability 2/6. The third book in 1 out of 5 ways with probability 1/5. The probability of making all three independent choices correctly is, therefore,

$$\left(\frac{3}{7}\right)\left(\frac{2}{6}\right)\left(\frac{1}{5}\right) = \frac{1}{35}$$

which agrees with the result of the first approach.

Example 2.3–4   *Probability of n Heads in 2n Tosses of a Coin*

The number of ways of getting exactly n successes in $2n$ equally likely trials is

$$C(2n, n)$$

and the probability of getting this is

$$C(2n, n)/2^{2n}$$

because the sum of all the binomial coefficients is 2^{2n}, by equation (2.3–6), (alternately each trial outcome has a probability of $1/2$).

To get an idea of this number we apply Stirling's formula (1.A–7). We get

$$(2n)!/\{n!n!2^{2n}\} \sim (2n)^{2n}e^{-2n}\sqrt{2\pi 2n}/\{n^n e^{-n}\sqrt{2\pi n}\,n^n e^{-n}\sqrt{2\pi n}\,2^{2n}\}$$
$$\sim 1/\sqrt{(\pi n)}$$

Thus the exact balancing of heads and tails in $2n$ tosses of a coin becomes increasingly unlikely as n increases—but slowly! At $n = 5$, (10 tosses), this approximation gives 0.24609 vs. the exact answer 0.25231, about 1 in 4 trials.

Example 2.3–5   *Probability of a Void in Bridge*

What is the probability when drawing 13 cards at random from a deck of 52 cards of having at least one suit missing?

For each success we must have drawn from one of 4 decks with only 39 cards (one suit missing). Hence the probability is

$$P = 4C(39, 13)/C(52, 13) = 4(39!)(39!)/(26!)(52!)$$
$$= 0.05116\ldots \sim 1/20$$

If your computer cannot handle a 52! then Stirling's approximation will yield $P = .05119\ldots$

Example 2.3–6   *Similar Items in Bins*

In how many ways can you put r indistinguishable items into n bins?

An example might be rolling r dice and counting how many have each possible face (are in each bin). The solution is simple once we realize that by adding to the r items $n - 1$ dividers ($|$) between bins; thus we are arranging $n + r - 1$ items

$$* * *\,|\,* *\,|\,* * * * *\,|\,* *\,|\,* * * * * * *\,|\,* \cdots *\,|\,* **$$

and this can be done in

$$C(n + r - 1, r) = C(n + r - 1, n - 1)$$

different ways.

Example 2.3–7   *With at Least 1 in Each Bin*

If in the above distribution we have to put at least one ball into each bin then of the r balls we distribute the first n into the n bins leaving $r - n$ balls and then proceed as before. This gives

$$C(r - 1, n - 1)$$

and we are assured of at least one in each bin.

Exercises 2.3

2.3–1   Write out the eleventh line of the Pascal triangle.

2.3–2   Use 2.3–2 to compute the eleventh line of the Pascal triangle.

2.3–3   What is the sum of the coefficients of the 11th line?

2.3–4   Compute $C(10, 5)$ directly.

2.3–5   Find $\sum_{k=1}^{11} kC(11, k)$.

2.3–6   Find $\sum k^2 C(n, k)$. Ans. $n(n + 1)2^{n-2}$.

2.3–7   Find an algebraic expression for $C(n, n - 1)$.

2.3–8   Show that $C(n, n - 2) = n(n - 1)/2$.

2.3–9   Show that the ratio of $C(n, n - 3)/C(n, n - 2) = (n - 2)/3$.

2.3–10   How many different bridge hands might you possibly get?

2.3–11   Discuss the fact that equation 2.3–1 always gives integers, that the indicated divisions always can be done.

2.3–12   For your machine find the comparative operation times and compute when the Pascal triangle is preferable to the direct computation of one line.

2.3–13   Show that if there are m Democrats and n Republicans then a committee consisting of k members from each party can be selected in $C(m, k) \times C(n, k) = n!m!/(n - k)! \, (m - k)!\{k!\}^2$ different ways.

2.3–14   Make a table of the probabilities of the number of heads in 6 tosses of a well balanced coin.

2.3–15   For 20 tosses $(n = 10)$ compare the Stirling approximation for 10 heads with the exact result. Ans. Exact $=.176197$, Est. $=.178412$.

2.3–16   What is the probability of a bridge hand having no cards other than 2, 3, 4, 5, 6, 7, 8, 9 and 10? Ans. $.003639. \ldots$

2.3–17   Find the sum of the terms of the form $k^3 C(n, k)$.

2.3–18   Compute $\sum_{k=0}^{n} C(n, k)/(n + 1)$.

2.3–19   If each of two people toss n coins what is the probability that both will have the same number of heads? (See 2.B–4) Ans. $C(2n, n)/2^{2n}$. Check this for $n = 1, 2, 3$.

2.3–20   In a deck of 52 cards one black card is removed. There are then 13 cards dealt and it is observed that all are the same color. Show that the probability that they are all red is 2/3.

2.3–21   If n items are put into m cells show that the expected number of empty cells is $(m - 1)^n/m^{n-1}$

2.3–22   If you expect 100 babies to be delivered in a hospital during the next 90 days, show that the expected number of days that the delivery room will not be in use is approximately 29.44 or about 1/3 of the time.

2.3–23   Balls are put into three cells until all are occupied. Give the distribution of the waiting time n. Ans. $(2^{n-1} - 2)/3^{n-1}, (n > 2)$.

2.3–24   Show that the probability of a hand in bridge of all the same color is $2C(26, 13)/C(52, 13) = 19/(47)(43)(41)(7) = 0.000032757. \ldots$

2.3–25   What is the approximate number of tosses of a coin when you can expect a 10% chance of half of the outcomes being heads?

2.4 The Binomial Distribution–Bernoulli Trials

Since the binomial distribution is so important we will repeat, in a slightly different form, much of the material just covered. Suppose that the probability of some event occurring is p, and of its not occurring is $q = 1 - p$. Consider n independent, repeated trials, called *Bernoulli trials*, in which there are exactly k successes, and of course $n - k$ failures. What is the probability of observing exactly k successes?

To begin we suppose that the first k trials are all successes and that the rest are all failures. The probability of this event is

$$p\,p\,p\ldots p\,q\,q\,q\ldots q = p^k q^{n-k}$$

Next, consider any other particular sequence of k successes whose positions in the run are fixed in advance, and $n - k$ failures in the remaining positions. When you pick that sequence of k p's and $(n - k)q$'s you will find that you have the same probability as in the first case.

Finally, we ask in how many ways the k successes and $(n - k)$ failures can occur in a total of n trials; the answer is, of course, $C(n, k)$. Hence when we add all these probabilities together, each having the same individual probability, we get

$$P(k) = C(n, k)p^k q^{n-k} \qquad\qquad (2.4\text{–}1)^*$$

as the probability of exactly k successes in n independent trials of probability p. This is often written as

$$b(k; n, p) = C(n, k)p^k q^{n-k} \qquad (0 \le k \le n) \qquad (2.4\text{–}2)^*$$

In words, "the binomial probability of k successes in n independent trials each of probability p." This gathers together all the $C(n, k)$ equally likely, $p^k q^{n-k}$, individual events in the original product sample space and groups them as one term. The result is the probability of exactly k successes in n independent trials, each single trial with probability of success p.

We now have the *probability distribution* for $b(k; n, p)$ as a function of the variable k. We see that this new distribution is not uniform. Even if $p = 1/2$ and the original sample space is uniform the grouped results are not.

There is a useful relationship between successive terms of this distribution which may be found as follows (from 2.3–2):

$$b(k + 1; n, p) = C(n, k + 1)p^{k+1}q^{n-k-1}$$

$$= \left(\frac{n - k}{k + 1}\right)\left(\frac{p}{q}\right)C(n, k)p^k q^{n-k} \qquad\qquad (2.4\text{–}3)$$

$$= \left(\frac{n - k}{k + 1}\right)\left(\frac{p}{q}\right)b(k; n, p)$$

With this we can easily compute the successive terms of the distribution.

For example, suppose $n = 10$, and $p = 2/3$, then as a function of k we have (using 2.4–3)

TABLE 2.4–1

$P(k)$	$= b(k; 10, 2/3)$
$P(0)$	$= 0.00002$
$P(1)$	$= 0.00034$
$P(2)$	$= 0.00305$
$P(3)$	$= 0.01626$
$P(4)$	$= 0.05690$
$P(5)$	$= 0.13658$
$P(6)$	$= 0.22761$
$P(7)$	$= 0.26012$
$P(8)$	$= 0.19509$
$P(9)$	$= 0.08671$
$P(10)$	$= 0.01734$
$Total$	$= 1.00000$

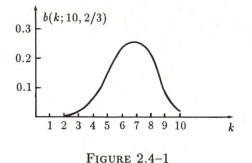

FIGURE 2.4–1

This is the *distribution* of the probability function $b(k; 10, 2/3)$, see Figure 2.4–1. The total probability must be 1, of course. To show that this is true *in general* we observe that the generating function of $b(k; n, p)$ can be found by

$$(q + pt)^n = \sum_{k=0}^{n} C(n, k)(pt)^k q^{n-k}$$

$$= \sum_{k=0}^{n} \{C(n, k)p^k q^{n-k}\} t^k \qquad\qquad (2.4\text{--}5)^*$$

$$= \sum_{k=0}^{n} b(k; n, p) t^k$$

where the coefficient of t^k is $b(k; n, p) = P(k)$. Now putting $t = 1$ we have

$$(q + p)^n = 1 = \sum_{k=0}^{n} b(k; n, p)$$

From the table we see that (within roundoff) the sum is indeed 1.

The sequence $b(k; n, p)$ is called the *binomial distribution* from its obvious source. It is also called the *Bernoulli distribution*; it arises whenever there are n independent binary (two way) choices, each having the same probability p of success, and you are interested in exactly k successes in the n trials.

For $p = 1/2$ the maximum of the binomial distribution is at the middle, $k = n/2$, (if n is even). The approximate inflection points of this important binomial distribution for $p = 1/2$, (which is a discrete distribution) can be found by setting the second difference approximately equal to 0 (see 2.3–1)

$$C(n, k + 1) - 2C(n, k) + C(n, k - 1)$$

$$= C(n, k) \left[\frac{n - k}{k + 1} - 2 + \frac{k}{n - k + 1} \right] \sim 0$$

Clearing the square bracket of fractions we have

$$n^2 - nk - nk + k^2 + n - k - 2(nk - k^2 + k + n - k + 1) + k^2 + k \sim 0$$

We arrange this in the form

$$4k^2 - 4nk + n^2 - n - 2 \sim 0$$

$$(2k - n)^2 \sim n + 2$$

$$(2.4\text{--}6)^*$$

$$k \sim \tfrac{1}{2}[n \pm \sqrt{(n + 2)}] \sim \frac{n}{2} \pm \frac{\sqrt{n}}{2}$$

and the inflection points for large n are symmetrically placed with respect to the position of the maximum, $n/2$, are at a distance approximately equal to $\sqrt{n}/2$.

Example 2.4–1   *Inspection of Parts*

If the probability of a defective part is 1/1000, what is the probability of exactly one defect in a shipment of 1000 items?

We reason as follows; the value from (2.4–1) is

$$b(1; 1000, 1/1000) = C(1000, 1)[(1/1000)(1 - 1/1000)^{999}]$$

$$= \tfrac{1000}{1000} (1 - 1/1000)^{1000}/(1 - 1/1000)$$

But remembering the limit from the calculus

$$\lim_{n \to \infty} (1 - 1/n)^n = 1/e$$

we apply this to the expression to get

$$(1/e)(1 - 1/1000) \sim 1/e \qquad\qquad (2.4-7)^*$$

This is a passable (not very good) approximation in many situations as can be seen from the following Table 2.4–2.

TABLE 2.4–2

n	$(1 - 1/n)^n$		Exact	Error in using $1/e$
1	0	=	0	
2	$(1/2)^2 = 1/4$	=	0.25	0.178794
3	$(2/3)^3 = 8/27$	=	0.296296	0.071583
4	$(3/4)^4 = 81/256$	=	0.316406	0.051473
5	$(4/5)^5 = 1024/3125$	=	0.327680	0.040199
10	$(9/10)^{10}$	=	0.348678	0.019201
20	$(19/20)^{20}$	=	0.358486	0.009393
50	$(49/50)^{50}$	=	0.364170	0.003709
100	$(99/100)^{100}$	=	0.366032	0.001847
200	$(199/200)^{200}$	=	0.366958	0.000921
500	$(499)/500)^{500}$	=	0.367511	0.000386
1000	$(999/1000)^{1000}$	=	0.367695	0.000184
2000	$(1999/2000)^{2000}$	=	0.367787	0.000092
5000	$(4999/5000)^{5000}$	=	0.367843	0.000036
10000	$(.9999)^{10,000}$	=	0.367861	0.000018

The limiting value is $1/e = 0.367879$. At $n = 10^k$ you get about k decimal places correct.

Example 2.4–2   *Continued*

Suppose the $p \ll 1$ (very much less) and $n \gg 1$. What is the probability of one or more defective parts in the sample?

This situation ("one or more") calls naturally for the complement probability approach and we set

$$P = 1 - Q$$

where $Q =$ probability of no defects $= (1 - p)^n$. Then

$$P = 1 - (1 - p)^n$$
$$= 1 - [(1 - p)^{1/p}]^{np}$$
$$\sim 1 - e^{-np}$$

If $np < 1$ then using the series expansion of $\exp(x)$

$$P \sim 1 - e^{-np} = 1 - [1 - np + (np)^2/2 - \cdots]$$
$$\sim np - (np)^2/2 + \cdots$$

The expected number of defects is np, and the next term is the first correction term for the multiple occurences.

Example 2.4–3   *Floppy Discs*

You are manufacturing floppy discs. Past experience indicates that about 1 in 10,000 discs that get out into the field are defective. Suddenly your processing and control systems change to about 1 in 100 defective. It is suggested that until the manufacturing process gets back into control you include in each package of 10 discs a note plus one extra disc, or maybe 2 extra discs. How effective do you estimate this to be?

Clearly the assumed probabilities are estimates and not exact numbers, hence we need only estimate things and do not need to use exact formulas. The ratio of the bad discs to the total in a pack is around $1/10$ in the new situation, and the first error term beyond what we are covering will give a valid estimate; we could find the exact computations by the "complement" approach if we thought it worth the trouble.

You were selling bad packages of 10 discs with a probability of about

$$1 - \text{probability that there are no bad discs in the 10}$$

This is, using the binomial expansion,

$$P = 1 - (1 - 1/10,000)^{10} \sim 1 - (1 - 10/10,000) \sim 1/1000$$

which is about the probability of one bad disc in 1000 packages

Since we will need estimates for various numbers we write out the general case,

$$\text{Prob } \{k \text{ bad discs}\} = b(k; n, p) = C(n, k)p^k(1 - p)^{n-k}$$

For 11 discs in a package, $n = 11$, we have the probability of two bad discs (with the new failure rate, $p = 1/100$)

$$C(11, 2)\{1/100^2\}\{1 - 1/100\} = 55\{99/100\}/10^4 \sim 5.4 \times 10^{-3}$$

which is about 5 times as bad as you were doing before the changes in the production line.

For 12 discs in a package the probability of 3 failures is

$$C(12, 3)p^3(1 - p)^9 \sim \{12 \times 11 \times 10/6\}10^{-6}\{1 - 9 \times 10^{-2}\}$$
$$= \{220/10^6\}\{0.91\} \sim 2 \times 10^{-4}$$

which is about 5 times better than you were doing!

Example 2.4–4   *The Distribution of the sum of Three Dice*

Sometime before the year 1642 Galileo was asked about the ratio of the probabilities of three dice having either a sum of 9 or else a sum of 10. We will go further and examine the whole probability distribution for the sum of three faces of three dice. Since the probability is not obvious we will use elementary methods.

We begin, as usual, with the sample space of the equally likely events; the roll of a single die at random means that we believe that each of the six faces has the same probability, 1/6. The second die makes the product space into 36 equally probable outcomes, each with probability 1/36. The third die leads to the product space of these 36 by its 6 giving $6^3 = 216$ events in the final product space, each with probability 1/216. Notice that the product space is the same whether we imagine the dice rolled one at a time or all at one time.

We do not want to write out all these 216 cases, rather we would like to get the sample space by a suitable grouping of events. If we label the sum (total value) of the three faces by S having values running from 3 to 18, then we want to find the probability distribution of S. The probability that S will have the value k is written as

$$Pr\{S = k\}   (k = 3, 4, \ldots, 18)$$

In our approach for fixed k we *partition* the value k into a sum of three integers, each in the range 1 to 6. We will first consider the *canonical partitions* where the partition values are monotonely increasing (or else monotonely decreasing). Once we have these we will then ask, "In how many places in the sample space will there be equivalent partitions?" We make the entries for the canonical partitions in the table on the right.

TABLE 2.4–3

Table of canonical partitions of k on three dice

k	canonical partitions						
3	(1,1,1)						1
4	(1,1,2)						3
5	(1,1,3)	(1,2,2)					6
6	(1,1,4)	(1,2,3)	(2,2,2)				10
7	(1,1,5)	(1,2,4)	(1,3,3)	(2,2,3)			15
8	(1,1,6)	(1,2,5)	(1,3,4)	(2,2,4)	(2,3,3)		21
9	(1,2,6)	(1,3,5)	(1,4,4)	(2,2,5)	(2,3,4)	(3,3,3)	25
10	(1,3,6)	(1,4,5)	(2,2,6)	(2,3,5)	(2,4,4)	(3,3,4)	27
11	(1,4,6)	(1,5,5)	(2,3,6)	(2,4,5)	(3,3,5)	(3,4,4)	27
12	(1,5,6)	(2,4,6)	(2,5,5)	(3,3,6)	(3,4,5)	(4,4,4)	25
13	(1,6,6)	(2,5,6)	(3,4,6)	(3,5,5)	(4,4,5)		21
14	(2,6,6)	(3,5,6)	(4,4,6)	(4,5,5)			15
15	(3,6,6)	(4,5,6)	(5,5,5)				10
16	(4,6,6)	(5,5,6)					6
17	(5,6,6)						3
18	(6,6,6)						1

total = 216

These are the increasing canonical partitions; in how many equivalent ways can each be written? If the three indices are distinct then there are evidently exactly 3! = 6 equivalent sequences in the entire sample space. If two indices are the same then the other index may be put in any of 3 places, hence there are 3 equivalent sequences in the sample space. Finally, if all three indices are the same then there is only one such sequence in the sample space. Thus we have to multiply each partition on the left by its multiplication factor (6, 3, or 1) and then sum across the line to get the total number of partitions that are in the original sample space and that also have the value k. These totals are given on the right. Dividing these sums by the total 216 we get the corresponding probabilities. When we notice the structure of the table, the symmetry of the totals above and below the middle, and check by adding all the numbers to see that we have not missed any, then we are reasonably confident that we have not made any mistakes.

The answer to the question asked of Galileo, the ratio of the probabilities of a sum of 9 or 10, is clearly $25/27 \sim 0.926$. Although there are the same number of canonical partitions in these two cases, the partitions do not have the same total number of representatives in the original sample space.

At the time of Galileo there were claims that the canonical partitions are the equally likely elements of the sample space. Hence *you* should review the argument we gave for the product sample space of probabilities to see if it convinces *you*.

The reader needs to be careful! The distribution we have used is known as the *Maxwell–Boltzmann distribution*. If we take the canonical partitions as the equally likely elements then the distribution is known as the *Bose–Einstein distribution*, which assumes that it is the entries on the left hand side of Table 2.4–1 that are the equally likely events and they have no corresponding multiplicative factors. Thus the right hand column would be the sequence (1, 1, 2, 3, 4, 5, 6, 6, 6, 6, 5, 4, 3, 2, 1, 1). The total number of equally likely cases is 56.

Finally, if we consider the Pauli exclusion principle of quantum mechanics then only the canonical partitions for which the three entries are distinct can occur and these are the equally likely events. We then have the *Fermi–Dirac distribution*. Thus in Table 2.4–1 we must eliminate all the entries for which two of the numbers are the same. When we do this we find, beginning with the sum 6 and going to the sum 15, sequence (1, 1, 2, 3, 3, 3, 3, 2, 1, 1). The total number of cases is 20.

Only the last two distributions, the Bose–Einstein and the Fermi–Dirac, are obeyed by the particles of physics. Thus you cannot argue solely from abstract mathematical principles as to which items are to be taken as the equally likely events; we must adopt the scientific approach and look at what reality indicates. If we wish to escape the medieval scholastic way of thinking then we must make a model, compute what to expect, and then verify that our model is (or is not) closely realized in practice (except when using "loaded dice").

Example 2.4–5   *Bose–Einstein Statistics*

If we put n indistinguishable balls in k cells then the number of ways we can do this is $C(n; j_1, j_2, \ldots j_k)$ equally likely ways, where the sum of the j_i is n. Hence any one configuration has the probability of the reciprocal of this number.

Example 2.4–6   *Fermi–Dirac Statistics*

Suppose we have k (indistinguishable) balls to put into n distinct boxes, ($k \leq n$), and ask, "In how many ways can this be done so that at most one ball goes into any one box?"

We can select the k places from the n possible ones in exactly

$$C(n, k)$$

ways, and each way is equally likely. Hence the probability of any one configuration is

$$1/C(n, k)$$

Exercises 2.4

2.4–1 Write out the table corresponding to (2.4–2) for the Bose–Einstein statistics.

2.4–2 Write out the table corresponding to (2.4–2) for the Fermi–Dirac statistics.

2.4–3 What is the probability of no defects in n items each having a probability of a defect p?

2.4–4 Using Exercise 2.4–3 what is the probability of two or more defective pieces?

2.4–5 Show that for large n $b(1; n, 1/n) \sim 1/e$.

2.4–6 Similarly show that $b(2; n, 1/n) \sim 1/2e$, and $b(3; n, 1/n) \sim 1/6e = 1/3!e$.

2.4–7 If p is the probability of a Bernoulli event and we make n trials show that the probability of an even number of events is $[1 + (2p - 1)^n]/2$.

2.4–8 Corresponding to Table 2.4–1 make a table of $b(k, 10, 3/5)$.

2.4–9 If a coin is biased with probability $p(H) = p$ then in the game of "odd man out" (see Exercise 1.7–15) what is the probability of a decision? Ans. $3pq$.

2.4–10 From an n-dimensional cube of lattice points and with a side n we select a random point. Show that the probability of the point being inside the cube is $[(n - 2)/n]^n \sim e^{-2}$.

2.4–11 Expand the binomials in the probabilities of 0, 1, 2, and 3 occurrences, and show that the expansions cancel out to the next term provided $np < 1$. Hence if $np \ll 1$ the first term neglected in the expansion is close to the exact result for 4 or more events.

2.4–12 Show by exact calculation that the floppy disc estimates in Example 2.4–3 are sufficiently accurate.

2.4–13 Find the distribution of the sum of two dice following the method used for three dice for the three distributions, Maxwell–Boltzmann, Bose–Einstein, and Fermi–Dirac.

2.4–14 If a coin has probabilty p of being a head show that the probability of k tosses of the coin will have the same side is $p^k + q^k$. For what p is this a minimum?

2.4–15 *Quality Control* In a set of 100 items 10 are defective. What is the probability that a sample of 10 will all be good? Ans. $\sim 1/e$.

2.4–16 Using the approximation $\exp(x) \sim (1 + 1/n)^{nx}$ and a similar one for $\exp(-x)$, discuss their product and what it indicates.

2.5   Random Variables, Mean and the Expected Value

We now introduce the idea of a *random variable*. We have, so far, discussed carefully the elementary events and the probabilities to assign to each. We now assign a *value* to the outcome of the event. For example, when we discussed the sum of the faces of three dice (Example 2.4–4) we had the value k associated with selected outcomes, namely where the sum of the three faces was equal to k. Thus we were assigning both a value 1, 2, ..., 6 to each of the corresponding faces of a die and also a value to the sum of the three faces. We wrote

$$Pr\{S = k\}$$

and we now understand S to be a random variable having values running from 3 to 18 with the corresponding probabilities $Pr\{S = k\}$. S is regarded as a function over the subsets of the sample space and

$$Pr\{S = k\}$$

is the sum of the probabilities of all outcomes which have the corresponding value k. This idea will be used extensively in the future, and if it is at first confusing this is the normal situation. The random variable idea arises because we assign values to the outcomes—the use of only names greatly restricts what we can compute.

Sometimes we want to know only a specific probability, but often we want to know the whole distribution for the random variable K which takes on the integer values k= $0, 1, \ldots, n$ with probabilities $P(k) = Pr\{K = k\}$ of exactly k successes in n trials (recall Table 2.4–1). In many cases, however, the actual numbers that make up the distribution are not easily assimilated by the human mind, and we need some summarizing descriptions of the distribution.

The first useful summarizing number is the *mean (average)*

$$\sum_{k=0}^{n} kp(k) = 0p(0) + 1p(1) + 2p(2) + \cdots + np(n) = \mu \qquad (2.5\text{–}1)^*$$

which is the mean value of the random variable K whose values are $k = 0, 1, \ldots, n$, each with the corresponding probability $p(k)$. This is usually labeled by the Greek lower case letter μ (mu), especially in statistics. It measures the position of the "center of the distribution." It is the mean value of the random variable K. It is also called the *average* or *expected value* of the random variable, or of the distribution.

Example 2.5–1   *The Expected (Mean) Value from the Roll of a Die*

The random roll of a die gives the faces 1, 2, 3, 4, 5, 6 each with probability 1/6. If we assign the *values* 1, 2, 3, 4, 5, 6 to the corresponding faces then the *expected* (average, mean) *value*

$$\mu = \tfrac{1}{6}[1 + 2 + 3 + 4 + 5 + 6] = \tfrac{1}{6}[6 \times 7/2] = 3.5$$

which is not a possible value from any roll. The expected value is not necessarily a value that can be "expected" to turn up! It is, at the moment, only a technical definition and not necessarily what a normal human would think! You have to be a very poor statistician to believe an expected value is a value you can always expect to see.

Example 2.5–2   *The Expected Value of a Coin Toss*

The random toss of a coin may be given the values: head = 1 and tail = 0. Thus the expected value of a toss $(p = 1/2)$ is

$$\mu = \tfrac{1}{2}[1 + 0] = \tfrac{1}{2}$$

However, we might have assigned the values: head = 1 and tail = −1. In this case the expected value would be

$$\mu = \tfrac{1}{2}[1 - 1] = 0$$

Which of these two assignments of values for the faces of the coin to use depends on the particular application—the assignment of values to the outcomes of the trials depends on the use you intend to make of the result. If you are counting the number of heads in a sequence of n tosses then the first assignment is reasonable and $\mu = 1/2$ is appropriate, but if you win one unit on a head and lose one unit on a tail then $\mu = 0$ is more reasonable; on a single toss you expect neither a gain nor a loss.

Example 2.5–3   *The Sum of Two Dice*

What is the expected value of the sum of the faces of two randomly thrown dice? We start with the sample space of $6 \times 6 = 36$ events each with probability of 1/36, (Exercise 1.6). We are interested in the sum so we now group together all events that have the same sum of the two faces, a value running from 2 to 12. This sum S is a random variable, say $S = X_1 + X_2$ where the X_1 is the random variable for the first die and X_2 is the random variable for the second die. We have the following Table 2.5–1:

TABLE 2.5–1

Table of $S = X_1 + X_2$

$P(S = 2)$	$= 1/36$	(1,1)					
$P(S = 3)$	$= 2/36$	(1,2)	(2,1)				
$P(S = 4)$	$= 3/36$	(1,3)	(2,2)	(3,1)			
$P(S = 5)$	$= 4/36$	(1,4)	(2,3)	(3,2)	(4,1)		
$P(S = 6)$	$= 5/36$	(1,5)	(2,4)	(3,3)	(4,2)	(5,1)	
$P(S = 7)$	$= 6/36$	(1,6)	(2,5)	(3,4)	(4,3)	(5,2)	(6,1)
$P(S = 8)$	$= 5/36$	(2,6)	(3,5)	(4,4)	(5,3)	(6,2)	
$P(S = 9)$	$= 4/36$	(3,6)	(4,5)	(5,4)	(6,3)		
$P(S = 10)$	$= 3/36$	(4,6)	(5,5)	(6,4)			
$P(S = 11)$	$= 2/36$	(5,6)	(6,5)				
$P(S = 12)$	$= 1/36$	(6,6)					

The table on the right lists the 36 equally likely entries in the sample space in a convenient order. The *derived* random variable S does not have a uniform distribution though it comes from a uniform distribution via grouping various terms having the same value for the random variable S (the sum). The expected value of the distribution of S is (for the standard value assignment to the outcomes)

$$\mu = [2 \times 1 + 3 \times 3 + 4 \times 3 + 5 \times 4 + 6 \times 5 + 7 \times 6 + 8 \times 5 + 9 \times 4$$
$$+ 10 \times 3 + 11 \times 2 + 12 \times 1]/36 = 252/36 = 7$$

But this is just the sum of the expected values from each of the two independently thrown dice! This is not a coincidence as we will see below. Thus further investigation of the expected value of a distribution (with a given value assignment) is worth some effort.

We see that the mean (average) or expected value is merely the sum of the values assigned to the outcomes of each trial when each outcome is weighted by the probability of its occurrence. This weighting of the values of the outcomes of the occurrences by their corresponding probabilities happens frequently and we need, therefore, a notation and some simple results to relieve us of rethinking the details every time we compute a mean. Given a random variable X whose outcomes have probabilities $p(i)$, and have the values x_i, $(i = 1, 2, \ldots, n)$, then the expected value of the random variable X is *defined* to be (E is for expectation)

$$E\{X\} = \sum_{i=1}^{n} x_i p(i) \qquad (2.5\text{--}2)^*$$

This is the average computed over the sample space, each outcome value x_i is weighted by its probability $p(i)$.

    The *expected value operator* $E\{.\}$ is a *linear operator* whose properties are worth investigating. We first examine the expected value of the product of two random independent variables X and Y, and then examine the sum.

    For the product XY, of the two random, independent variables X and Y, the probability of the pair of outcomes x_i and y_j will by definition (2.5–2) be

$$E\{XY\} = \sum_{k} k p\{XY = k\} \qquad (2.5\text{--}3)$$

How do we get the $p\{XY = k\}$? By definition (2.5–2) we must first sum all the $p_X(i) p_Y(j)$ in the sample space for which $x_i y_j = k$, where the subscripts on the probabilities indicate the corresponding variable. Hence we have

$$E\{XY\} = \sum_{k} k \sum_{x_i y_j = k} p_X(i) p_Y(j)$$

    This formula is not easy to use so we seek an alternate approach to compute $E\{XY\}$. We go back to (2.5–2). The product $x_i y_j$ arises with probability $p_X(i) p_Y(j)$, hence

$$E\{XY\} = \sum_{\substack{\text{sample} \\ \text{space}}} x_i y_j p_X(i) p_Y(j)$$

$$= \sum_{i} \sum_{j} x_i y_j p_X(i) p_Y(j) \qquad (2.5\text{--}4)$$

We need to convince ourselves that these two different expressions (2.5–3) and (2.5–4) for $E\{X\}$ are the same. To see this take any point in the product sample space (x_i, y_j) and follow it through each computation; you will see that it enters into the result exactly once and only once either way, each time with weight $p_X(i) p_Y(j)$. Hence the two formulas represent alternate ways of finding the expected value.

    This proof of the equivalence is simple for a finite discrete sample space, but when we later face continuous sample spaces it brings up two separate ideas. First, the integral (corresponding to the finite summation) over the whole sample space is the iterated integral (in either order) as in the calculus. Second, a change of variables (in this case from rectangluar to what will be an integration along a hyperbolic coordinate and then all the hyperbolas are integrated to cover the whole sample space) to new coordinates will involve the corresponding Jacobian of the transformation.

We now return to the problem at hand. Taking it in the second form, (2.5–4), we have the expected value of the product of the independent variables

$$E\{XY\} = \sum_{i,j} x_i y_j p_X(i) p_Y(j)$$

If we fix in our minds one of the variables, say x_i, and sum over the y variable we will find that for *each* value x_i we will get the expected value of Y, namely $E\{Y\}$. This will factor out of each term in the variable x_i, hence out of the sum. What is left is the expected value of X. Thus we have the result that for independent random variables

$$E\{XY\} = \sum_i x_i p_X(i) \sum_j y_j p_Y(j)$$

$$= E\{X\}E\{Y\}$$

(2.5–5)

A rather simple way to see this important result is to actually write out the product space of the random independent variables X and Y in a rectangular array.

$$\begin{vmatrix} x_1 y_1 p_X(1) p_Y(1) & x_1 y_2 p_X(1) p_Y(2) & \cdots & x_1 y_n p_X(1) p_Y(n) \\ x_2 y_1 p_X(2) p_Y(1) & x_2 y_2 p_X(2) p_Y(2) & \cdots & x_2 y_n p_X(2) p_Y(n) \\ \cdots & \cdots & & \cdots \\ x_n y_1 p_X(n) p_Y(1) & x_n y_2 p_X(n) p_Y(2) & \cdots & x_n y_n p_X(n) p_Y(n) \end{vmatrix}$$

Now consider summing by the rows; for the ith row the $x_i p_X(i)$ value is fixed and you get for every row the same value $E\{Y\}$. Now sum what is left after factoring out the $E\{Y\}$ and you get $E\{X\}$. Alternately, you can sum by the columns first to get $E\{X\}$ and then sum the rest to get the $E\{Y\}$. In both cases you get the product of the expectations.

Next we consider the sum of any two random variables X and Y, *independent or not*. We have

$$E\{X + Y\} = \sum_i^j (x_i + y_j) p(i,j)$$

where $p(i,j)$ is the probability of the pair x_i, y_j occurring. When you think about the two following equations you see why they are true:

$$\sum_j p(i,j) = p_X(i)$$

$$\sum_i p(i,j) = p_Y(j)$$

(2.5–6)

We have, therefore, on breaking up the sum into two sums and then interchanging the order of summation in the second sum,

$$E\{X+Y\} = \sum_i^j (x_i + y_j)p(i,j)$$

$$= \sum_i x_i \sum_j p(i,j) + \sum_j y_j \sum_i p(i,j)$$

$$= \sum_i x_i p_X(i) + \sum_j y_j p_Y(j)$$

$$= E\{X\} + E\{Y\}$$

This should be clearly understood because it is fundamental in many situations. We have the result that the expected value of the sum of two random variables, *independent or not*, is the sum of their expected values,

$$E\{X+Y\} = E\{X\} + E\{Y\} \qquad (2.5\text{–}7)$$

It is an easy extension to see that $E\{.\}$ is a *linear operator*, that is

$$E\{aX + bY\} = aE\{X\} + bE\{Y\} \qquad (2.5\text{–}8)^*$$

We have only to follow through the above argument with the constants a and b and notice how they factor out of each sum. This result shows why the expected value of the sum of the values of the two dice is the sum of their expected values.

The fact that the expectation of a sum of random variables is the sum of the expectations *no matter how the two variables are related* explains why $E\{.\}$ is such a useful *linear operator*.

This method of summing over one of the variables and then over the other, and finding that the expected value factors out of the other sum and that then the other sum is exactly 1, is fundamental to much of the theory and should be mastered. If you do not see it clearly then look again at the product space (the rectangular array) and examine what happens as you sum $(x_i + y_j)$ by rows and then by columns.

Once we notice that the sum of two random variables is itself a random variable, then we can recursively apply the above formula and obtain result that for any finite sum of random variables X_i with constants c_i

$$E\{\sum c_i X_i\} = \sum c_i E\{X_i\} \qquad (2.5\text{–}9)^*$$

For products of *independent* random variables we have correspondingly

$$E\{\prod_i [X_i]\} = \prod_i [E\{X_i\}] \qquad (2.5\text{–}10)^*$$

these generalize equations (2.5–7) and (2.5–5).

Other measures of a distribution besides the expected value are often used by statisticians. One is the *median* which is the middle value when the x_i are ranked in order (the average of the middle two when there are an even number of values). Another measure is the *mode* which is the most frequent (most fashionable) value. The mode makes sense *only* when the distribution has a single, well defined, peak. The mid range $(x_{max} - x_{min})/2$ is occasionally used.

Example 2.5–4   *Gambling on an Unknown Bias*

You have a sequence of repeated, independent trials of unknown probability p, say betting on a biased coin. One strategy for betting is to bet on the face of the most recent outcome thus making your bets have the same frequency as the out comes of the trials. Your expected gain per trial is

$$\text{gain} - \text{loss} = pp + qq - \{pq + qp\} = (p - q)^2$$

If you have the additional information that the bias is such that $p > 1/2$ then an alternate strategy is to bet always on the event. Now your gain per toss is

$$(p - q)$$

Thus knowing only the side which the bias favors, you have a significant advantage (except for $p = 1/2$ and $p = 1$) as the following table shows.

prob. p	gain 1	gain 2
.5	.00	.0
.6	.04	.2
.7	.09	.4
.8	.16	.6
.9	.64	.8
1.0	1.00	1.0

Mathematical aside: *Sums of Powers of the Integers*

We often need some small mathematical details that are not easily recalled by the student. In particular we will often need the sums of the consecutive integers raised to low powers; we will need the formulas

$$\sum_{k=1}^{n} k = n(n+1)/2$$

$$\sum_{k=1}^{n} k^2 = n(n+1)(2n+1)/6$$

$$\sum_{k=1}^{n} k^3 = [n(n+1)/2]^2$$

These are all easy to prove by the use of mathematical induction. For example, for the sum of the cubes of successive integers we need to show: (1) for the basis of the induction, using $n = 1$, we get the same number on both sides, and (2) for the induction step from $n - 1$ to n we have an equality, namely

$$\text{old sum} + \text{next term} = \text{new sum}$$

$$[n(n - 1)/2]^2 + n^3 = n^2[(n^2 - 2n + 1)/4 + n]$$

$$= n^2[(n + 1)^2/4] = [n(n + 1)/2]^2 = \text{new sum}$$

Exercises 2.5

2.5–1   Cannon balls are piled as a pyramid with a square base. If the side of the square is n, then how many balls in the pile?

2.5–2   Triangular numbers are defined as $f(n) = n(n - 1)/2$. Find the sum of the first k triangular numbers.

2.5–3   What is the expected value for a deck of cards numbered (values) 1, 2, ..., 52?

2.5–4   What is the expected value of the sum of four dice?

2.5–5   What is the expected value of the sum of n coins? Ans. $n/2$ or 0.

2.5–6   What is the expected value of the sum of n dice?

2.5–7   One and only one key on a key chain of n keys fits the lock. Compare finding the key by systematic or random methods.

2.5–8   A disc track is selected from a set numbered 1, 2, ..., n. Call this X. You now select Y from $1 \leq y \leq X$. Show that $E\{Y\} = (n + 3)/4$.

2.5–9   If $p(k) = k/10$, $(0 \leq k \leq 4)$ find $E(K)$.

2.5–10   If $p(k) = qp^k (k = 0, 1, ...)$ find $E(K)$.

2.5–11   Find the product of the values on the two faces of a pair of independently thrown dice. Ans. 49/4.

2.5–12   What is the expected value of the product of three independently thrown dice?

2.6 The Variance

The mean gives the location of the "center" of a distribution, but a little experience with distributions soon shows that some distributions are spread out and some are narrow; see Figure 2.6–1. The amount of the spread is often important so we need a measure of the "spread" of a distribution. The most

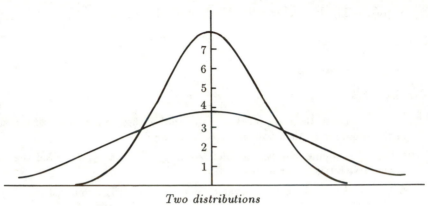

Two distributions

FIGURE 2.6–1

useful one *from the point of view of mathematics* (meaning it can be handled easily) is the sum of the squares of the deviations from the mean, μ, each square of course multiplied by its probability $p(i)$,

$$\text{Variance } \{X\} = V\{X\} = \sum_i (x_i - \mu)^2 p(i) = \sigma^2 \qquad (2.6\text{–}1)^*$$

The standard notation is σ^2, where σ is often used as a measure for the spread of the distribution of the random variable X. A large variance means that the distribution is broad, and a small variance means that the all distribution is near the mean. Indeed, a zero variance means that the distribution is all at one point, the mean, μ. Note, like the mean, the variance is computed over the whole sample space.

It is immediately evident that *the variance is independent of any shift in the coordinate axis*, since it depends only on the difference of coordinates and not on the coordinates themselves. This is important because it allows us, many times, to simplify various derivations by assuming that the mean is zero.

We need the facts that (c is a constant)

$$V\{c\} = 0 \qquad V\{cX\} = c^2 V\{X\} \qquad (2.6\text{–}2)^*$$

The first follows because the variance is measured about the mean (which is c), and the second because by the definition a multiplying factor c will also multiply the mean by c, and hence c^2 comes out of the square of the differences.

For theoretical work we often convert the formula for the variance to a more useful form. We simply expand the square term in (2.6–1) and note how the mean arises, ($E\{X\} = \mu$)

$$V\{X\} = E\{X^2\} - 2\mu E\{X\} + \mu^2 = E\{X^2\} - E^2\{X\} \qquad (2.6\text{–}3)^*$$

Next, we prove that for a sum of *independent* random variables X_i we have ($V\{.\}$ is a linear operator)

$$V\left\{\sum X_i\right\} = \sum V\{X_i\} \qquad (2.6\text{–}4)^*$$

The proof is easy. For convenience we assume that the mean of each random variable X_i is 0, $E\{X_i\} = 0$. Then we have, (using different summation indices to keep things straight)

$$E\left\{\sum X_i^2\right\} = E\left\{\sum X_i \sum X_j\right\}$$

$$= E\left\{\sum_{i,j} X_i X_j\right\}$$

$$= E\left\{\sum_i X_i^2\right\} + E\left\{\sum_{i \neq j} X_i X_j\right\}$$

where the squared terms are in the first summation and the cross products are in the second (each there twice). But for independent random variables from (2.5–5) $E\{X_i X_j\} = E\{X_i\}E\{X_j\} = 0 \times 0 = 0$ (since we assumed the means were 0). Hence the last sum is zero. Therefore, *for independent random variables* the variances simply add, a very useful property:

$$V\left\{\sum X_i\right\} = \sum V\{X_i\} \qquad (2.6\text{–}5)^*$$

Thus the variance of the sum of n samples of the same random variable has the variance $n\sigma^2$.

The mean and variance we have just computed are probability weighted averages of the values over the whole sample space. Often for just n random sample values from the distribution we want to know *expected value and the variance of their average*. Thus we need to study the average of the n samples. We do not want to actual handle the specific samples, so we write their corresponding random variables X_i, and examine the random variable

$$S(n) = X_1 + X_2 + \cdots + X_n$$

We want to study the average so we use

$$S(n)/n = [X_1 + X_2 + \cdots + X_n]/n$$

For the expected value of the average $S(n)/n$ we have

$$E\{S(n)/n\} = [E\{X_1\} + \cdots + E\{X_n\}]/n$$
$$= [\mu + \mu + \mu + \cdots + \mu]/n = \mu$$

Thus the expected value of the average of n samples is exactly the expected value of the original random variable X—it is an *unbiased estimate* since it is neither too high nor too low.

For the variance of the average of n independent samples we assume the mean is 0 and have

$$V\{S(n)/n\} = [V\{X_1\} + \cdots + V\{X_n\}]/n^2$$
$$= \sigma^2/n$$

Thus the variance of the average of n samples of a random variable decreases like $1/n$. The corresponding measure of the spread of the distribution (compare 2.4–6) of the average of n samples is

$$\sigma/\sqrt{n}$$

This means that if we want to decrease the spread of a distribution of the average of n samples by a factor of 10 we will need to take 100 times as many samples.

How robust are these measures to small changes in the probabilities $p(i)$? If the $p(i)$ are not exactly what we supposed they were but have errors $e(i)$, then the true values are

$$p(i) + e(i)$$

where we have, of course, $\sum e(i) = 0$. We should have computed

$$E\{X\}^* = \sum x_i[(p(i) + e(i)] = E\{X\} + \sum x_i e(i)$$

and we lost the last sum. For convenience we write

$$E\{X\}^* = \mu^* \quad \text{and} \quad E\{X\} = \mu$$

and we have

$$\mu^* = \mu + \sum x_i e(i) = \mu + A \quad \text{or} \quad \mu^* - \mu = A$$

where we have set $A = \sum_i x_i e(i)$ and A is therefore the shift in the mean. The variance is more complex. We should have computed

$$V^* = \sum(x_i - \mu^*)^2 (p(i) + e(i))$$

Instead we computed

$$V = \sum(x_i - \mu)^2 p(i)$$

To get V^* in terms of simpler expressions, especially V, we write things as

$$V^* = \sum(x_i - \mu - A)^2 p(i) + \sum(x_i - \mu - A)^2 e(i)$$

Expand the binomials and get term by term

$$V^* = V - 2A \sum(x_i - \mu)p(i) + A^2$$
$$+ \sum(x_i - \mu)^2 e(i) - 2A \sum(x_i - \mu)e(i) + A^2 \sum e(i)$$

The second and sixth terms and part of the fourth and fifth drop out and if we set $\sum x_i^2 e(i) = B$ then we have

$$V^* = V + A^2 + B - 2\mu A - 2A^2$$

$$= V + B - A(2\mu + A)$$

Example 2.6–1   The Variance of a Die

The roll X of a die has, (from Example 2.5–1), $E\{X\} = 7/2$. The variance is the values measured from the mean, then squared, and finally summed over all the values weighted by their probabilities $p(i)$. We get, for the standard assignment of values for the faces of the die, the differences from the mean

$$-5/2, \quad -3/2, \quad -1/2, \quad 1/2, \quad 3/2, \quad 5/2$$

to be squared, multiplied each by $p(i) = 1/6$, and added. As a result we get

$$(1/6)[25 + 9 + 1 + 1 + 9 + 25]/4 = 70/24 = 35/12$$

Instead we can compute the squares of 1, 2, 3, 4, 5, 6, add, and then multiply the sum by the probability 1/6, to get $E\{X^2\} = 91/6$. According to (2.6–3) we then subtract the square of the mean

$$91/6 - (7/2)^2 = 91/6 - 49/4 = (182 - 147)/12 = 35/12$$

which is, of course, the same number.

Notice that the formula (2.6–3) may (as in this example) give a difference of large numbers and hence be sensitive to roundoff errors, especially when the mean is large with respect to the variance.

Example 2.6–2   *Variance of M Equally Likely Outcomes*

This is a generalization of Example 2.6–1. We have the $x_i = i$, $(i = 1, \ldots, M)$, $p(i) = 1/M$, and the sums

$$\sum 1 = M$$

$$E\{X\} = (1/M) \sum x_i = (1/M)M(M+1)/2 = (M+1)/2$$

$$E\{X^2\} = (1/M) \sum x_i^2 = (1/M)M(M+1)(2M+1)/6$$

$$= (M+1)(2M+1)/6$$

hence the variance is

$$V = [(M+1)][(2M+1)/6 - (M+1)/4]$$

$$= (M+1)[4M+2-3M-3]/12$$

$$= (M+1)(M-1)/12 = (M^2-1)/12$$

For $M = 6$ this gives $35/12$ as before, Example 2.6–1.

Example 2.6–3   *The Variance of the Sum of n Dice*

If we assume, as we should since nothing else is said, that the outcomes of the roll of n individual dice are independent and uniformly likely, then by (2.6–5) we merely take the variance of a single die and multiply by n

$$(35/12)n$$

to get the variance of the sum of n dice.

There is a tendency to believe that the square root of the variance gives a good measure of the expected deviation from the mean. But remember that we "averaged" the squares of the deviations from the mean. In the sum of the squares the big deviations tend to dominate the sum (since the square of a large number is very large and the square of a small number is small, hence only the large numbers influence the total very much). Thus it should be clear that if the deviations from the mean vary a lot then the square root of the variance is *not* a good measure of the *expected deviation*, which should be given by a formula like (although μ should probably be computed by a different formula, say the median)

$$E|X-\mu| = \sum_i |x_i - \mu| p(i)$$

This is the average of the absolute differences from the mean (or maybe the median). It is not used much because it has difficult mathematical properties (but often good statistical properties!).

Consider the equally likely data $(p(i) = 1/10)$

$$x_1 = 6, x_2 = x_3 = \cdots = x_{10} = 1$$

The mean is $(6 + 9)/10 = 1.5$, hence the variance is

$$\text{Var} = (1/10)[(4.5)^2 + 9(0.5)^2] = (20.25 + 2.25)/10 = 2.25$$

Hence the *root mean square* (the square root of the mean of the quares) is 1.5. On the other hand the mean deviation from 1.5 (the mean) is

$$(1/10) \sum \mid x_i - 1.5 \mid = (1/10)[4.5 + 9(1/2)] = 9/10 = 0.9$$

The difference in the two answers is due mainly to the "outlier" $x_1 = 6$. When is 1.5 a better measure of the deviation from a reasonable estimate of the center of the distribution than is 0.9 given nine values at 1 and one value at 6? It all depends on the use to be made of the result! Had we used the median, which is 1, and found the mean deviation from it we would, in this case, have found the result $1/2$.

The sequence $E\{1\} = 1, E\{X\}$, and $E\{X^2\}$ suggests that the general *kth moment* should be defined by

$$E\{X^k\} = \sum_i x_i^k p(i) \qquad (2.6\text{–}5)^*$$

If we want the kth *moment about the mean* (the kth *central moment*) then we have

$$E\{(X - \mu)^k\} = \sum_i (x_i - \mu)^k p(i) \qquad (2.6\text{–}6)^*$$

For most purposes all the moments determine, theoretically, the distribution, and play a significant role in both probability and statistics. In practice only the lower moments are used—because as noted above the larger terms dominate in the sum of the squares and this is even more true for higher powers. The third moment about the mean is called "skewness" and the fourth is called "kurtosis" (elongation, flatness); the higher ones, have not even been named! Similarly,

$$E\{f(n)\} = \sum_i f(i)p(i) \qquad (2.6\text{–}7)^*$$

for any reasonable function $f(n)$.

Exercises 2.6

2.6–1 For the formula (2.6–5) find for $k = 3$, 4, and 5 the central moments in terms of the moments about the origin.

2.6–2 Carry out the derivation of the formula for the variance when the mean is μ.

2.6–3 Find the variance of n trials from a sample space of M equally likely events.

2.6–4 Conjecture that the sum of the fourth powers of the integers is a polynomial in n of degree five. Impose the induction hypothesis (and initial sum) and deduce the corresponding coefficients of the fifth degree polynomial.

2.6–5 Find the mean and variance of a Bernoulli distribution.

2.6–6 Find the third and fourth moments of a Bernoulli distribution.

2.6–7 Find the mean and variance of the distribution $\{qp^n\}$ $(n = 0, 1, \ldots)$. Ans. p and p/q^2.

2.6–8 For the distribution $p(k) = k/10$ $(k = 0, \ldots, 4)$ find the mean and variance.

2.7 The Generating Function

Given a probability distribution $p(k)$ (for integer values k) we can write it as a single expression by means of the *generating function* which is a polynomial in the variable t whose coefficients are the values $p(k)$

$$G(t) = \sum_k p(k)t^k = E\{t^k\} \qquad (2.7\text{–}1)^*$$

For example, for a die with six faces each of probability 1/6, we have the generating function

$$G(t) = (t + t^2 + t^3 + t^4 + t^5 + t^6)/6$$

$$= t(1 - t^6)/6(1 - t)$$

We have seen the generating function before, but in the form used for counting or enumerating. For example, for the binomial coefficients we had (2.3–2) the generating function

$$G(t) = (1 + t)^n = \sum_{k=0}^{n} C(n, k)t^k$$

This suggests immediately that for the binomial (Bernoulli) probability distribution (2.4–2)

$$b(k; n, p) = C(n, k)p^k q^{n-k}$$

we will have the generating function (see 2.4–5)

$$G(t) = \sum_{k=0}^{n} \{C(n, k)p^k q^{n-k}\}t^k = (q + pt)^n \tag{2.7-2}$$

The generating function representation of a probability distribution is very useful in many situations; *the whole probability distribution is contained in a single expression.* It should be evident from (2.7–1) that

$$G(1) = \sum_{k} p(k) = 1$$

If we differentiate the generating function with respect to t we get

$$G'(t) = \sum_{k} kp(k)t^{k-1}$$

and hence

$$G'(1) = \mu = E\{X\} \tag{2.7-3}^*$$

If we multiply $G'(t)$ by t and then differentiate again we will get

$$[tG'(t)]' = tG''(t) + G'(t) = \sum_{k} k^2 p(k)t^{k-1}$$

and hence we have (set $t = 1$ again)

$$E\{X^2\} = G''(1) + G'(1)$$

Therefore, the second moment about the mean is simply (2.6–3)

$$V\{X\} = G''(1) + G'(1) - [G'(1)]^2 = \sigma^2 \tag{2.7-4}^*$$

We see that *from the generating function alone* we can get by formal differentiation both the mean and variance of a distribution (as well as higher moments if we wish). Thus the generating function is a fundamental tool in handling distributions and solving probability problems.

Example 2.7–1   The Mean and Variance of the Binomial Distribution

From (2.7–2) the generating function for the binomial distribution and its first two derivatives are

$$G(t) = (q + pt)^n$$

$$G'(t) = np(q + pt)^{n-1}$$

$$G''(t) = n(n - 1)p^2(q + pt)^{n-2}$$

We now set $t = 1$ and get the values (remember that $p + q = 1$)

$$G(1) = 1$$

$$G'(1) = \mu = np$$

$$G''(1) = n(n - 1)p^2$$

hence we have for the binomial distribution (use (2.7–4))

$$\text{mean } = \mu = np \tag{2.7–5}^*$$

$$\text{variance } = \sigma^2 = n(n - 1)p^2 + (np) - (np)^2$$

$$= np(1 - p) = npq \tag{2.7–6}^*$$

The generating function allows us to do much more. If we ask how the sum of the faces of two dice can arise we see that it is the product of the generating function with itself

$$G(t)G(t) = (t + t^2 + \cdots + t^6)(t + t^2 + \cdots + t^6)$$

$$= t^2 + 2t^3 + 3t^4 + \cdots + t^{12}$$

The coefficient of any power of t, say the kth power, is the number of ways that the sums of the original exponents can produce that k. The number of ways this can happen is exactly the coefficient of t^k. The coefficient *counts* the number of ways the sum can arise. Thus we have the result that the generating function for the sum of the faces of two dice is simply the square of the generating function for a single die.

This is a general result. Since the exponent of t is the same as the value assigned to the outcome of the trial, the power series product of two generating functions (where the exponents are automatically arranged for us when we arrange things in powers of t) has for each power of t a coefficient that counts how often that sum can occur. In general, if

$$G(t) = \sum_i a_i t^i$$

$$H(t) = \sum_j b_j t^j$$

(3.7–8)

where the summations begin with index 0 to allow for a possible constant term), then for the product we have

$$G(t)H(t) = \sum_k c_k t^k \qquad (3.7–9)^*$$

where

$$c_k = \sum_{i+j=k} a_i b_j = \sum_{i=0}^{k} a_i b_{k-i} \qquad (3.7–10)^*$$

This sequence $\{c_k\}$ is called *the convolution* of the sequence $\{a_i\}$ with the sequence $\{b_j\}$.

The concept of a convolution is very useful, so we will linger over it a bit more to improve your understanding of it. We can easily picture a convolution of two sequences, one belonging to $G(t)$ and the other to $H(t)$. We imagine one written from left to right on a strip, beginning with the zeroth term, and the other sequence written in the reverse direction (right to left) on the other

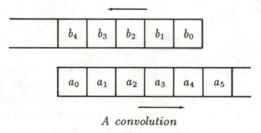

A convolution

FIGURE 2.7–1

strip, as shown in Figure 2.7–1. The second is placed below the first and is also displaced so that the kth term of one is opposite the zeroth term of the other. The sum of the products of the overlapping terms is exactly the convolution of the two corresponding sequences. In more detail we start with the zeroth terms opposite each other, compute to get the value for $c_0 = a_0 b_0$; shift one strip one unit, compute to get the value for $c_1 = a_0 b_1 + a_1 b_0$; shift, compute $c_2 = a_0 b_2 + a_1 b_1 + a_2 b_0$; ...until we get all the coefficients c_k we want. The convolution of $G(t)$ with $H(t)$ is the same as the convolution of

$H(t)$ with $G(t)$ as can be seen from the figure or else from the formula (3.7–10) by replacing the dummy summation index i by $k - i$. Hence the convolution operation is commutative {it must be since $G(t)H(t) = H(t)G(t)$}.

Example 2.7–2   *The Binomial Distribution Again*

With the idea of a convolution in mind we start with the simple observation that a single binomial choice with probabilities p and q has the generating function

$$G(t) = (q + pt)$$

hence the sum of n independent binomial choices will have this same generating function raised to the nth power, (2.7–2),

$$G^n(t) = (q + pt)^n$$

as we should. From the expression for $G(t) = (q + pt)$ we see that $G'(1) = p$ and $G''(1) = 0$, hence using (2.5–6) and (2.6–3) we have for the binomial distribution of order n

$$\mu = np$$

$$\text{Var} = n(0 + p - p^2) = np(1 - p) = npq$$

as before, Example 2.7–1.

Example 2.7–3   *The Distribution of the sum of Three Dice*

At the start of this Section we gave the generating function of the face values of the roll of a single die,

$$G(t) = t(1 - t^6)/6(1 - t)$$

To get the generating function for the roll of 3 dice we merely cube this. It is not going to be easy to get the coefficients of this generating function by a power series expansion, (but see Example 2.4–4). Instead we can use the idea of a convolution and observe that the roll of 2 dice is the convolution of the sequence with itself (neglecting the factor 1/6), where the "|" is at the beginning of the sequence (just before the 0th term) for the first die

$$|0, \ 1, \ 1, \ 1, \ 1, \ 1, \ 1, \ 0, \ 0, \ 0 \ldots$$

We have supplied 0 coefficients as needed. Convolving this with itself (remember to reverse the order of the coefficients on the second strip) we see that we get, neglecting the factor 36 for the moment and merely counting the number of times,

$$|0, \ 0, \ 1, \ 2, \ 3, \ 4, \ 5, \ 6, \ 5, \ 4, \ 3, \ 2, \ 1, \ 0, \ 0, \ \ldots$$

To get the result for three dice we convolve this with the original sequence and get,

$$|0, \ 0, \ 0, \ 1, \ 3, \ 6, \ 10, \ 15, \ 21, \ 25, \ 27, \ 27, \ 25, \ 21, \ 15, \ 10, \ldots$$

as we did in Table 2.4–1. To get the probabilities we divide by $6^3 = 216$. A comparison of the details of the two methods of computing shows that we are making progress when we introduce the more advanced ideas of generating functions and convolution—they help make actual computation easy as well as making the thinking a lot easier.

Two things should be noted about our definition of a generating function in terms of a probability distribution. In the first place, given any sequence of numbers a_i we can define a generating function

$$A(t) = \sum_k a_k t^k$$

The derivatives at the value $t = 1$ are

$$A'(1) = \sum_k k a_k$$

$$A''(1) = \sum_k k(k-1) a_k$$

There is an alternate form of a generating function that should perhaps be mentioned. The *exponential generating function* is defined by

$$M(t) = \sum_k a_k t^k / k!$$

For some purposes the $M(t)$ is useful, but the simple formulas for the mean and variance require modification.

Exercises 2.7

2.7–1 Find the generating function, mean and variance for four dice.

2.7–2 Same for two dice with one biased.

2.7–3 Same with both dice biased different amounts.

2.7–4 Apply the convolution theorem to the Bernoulli distribution and deduce various identities among the $b(k; n, p)$.

2.7–5 Apply the generating function approach to the toss of coins. Show that for n coins you get the correct mean and variance.

2.7–6 If the generating function $G(t)$ is the kth power of another generating function $F(t)$ find the first and second derivatives of $G(t)$.

2.7–7   You draw one spade and one heart. What is the distribution of the sum of the faces using values 1, 2, ..., 13 if you draw one card from each of these suits?

2.7–8   Find the distribution of a biased die rolled twice.

2.7–9   A distribution has p_0, p_1 and p_2 (with of course their sum being 1). Find the distribution of the sum of two and of three trials. Ans. For two trials, p_0^2, $2p_0p_1$, $2p_0p_2 + p_1p_2^2$.

2.7–10   For the distribution $\{qp^n\}$ find the distribution of the sum of two and of three events happening.

2.8   The Weak Law of Large Numbers

We now examine one of the most important results in the simple theory of probability, the *weak law of large numbers*. It answers the question, "What is the expected value for the average of n independent samples of the same random variable X?" This law connects our postulated idea of probability, which is based on symmetry, with our intuitive feeling that the average approaches, in some sense, the expected value—that the formally defined expected value $E\{X\}$ is actually the "expected value" from many trials. It is called the "weak law" because there is a stronger form of it which uses the limit as the number of samples approaches infinity.

Before proving this important result we need a result called the *Chebyshev inequality*. Given a random variable X whose values, positive or negative, are the x_i with corresponding probabilities $p(i)$, we form the sum

$$E\{X^2\} = \sum_i x_i^2 p(i)$$

From this sum we exclude those values which are within a distance ϵ (Greek lower case epsilon) of the origin. Hence we have the inequality

$$E\{X^2\} \geq \sum_{|x_i| \geq \epsilon} x_i^2 p(i)$$

We next strengthen the inequality by reducing all the x_i^2 to the smallest value they can have, namely ϵ^2. We have

$$E\{X^2\} \geq \epsilon^2 \sum_{|x| \geq \epsilon} p(i)$$

and on writing Pr{.} as the probability of the condition in the {.} occurring

$$E\{X^2\} \geq \epsilon^2 \Pr\{|x_i| \geq \epsilon\}$$

This can be rearranged in the form

$$\Pr\{|x_i| \geq \epsilon\} \leq E\{X^2\}/\epsilon^2 \qquad (2.8\text{--}1)^*$$

which is known as the *Chebyshev inequality*.

There is another form of the Chebyshev inequality that is often more useful for thinking; this measures the deviation in terms of the quantity σ. We merely replace ϵ by $\sigma\epsilon$. This gives us, corresponding to (2.8–1)

$$\Pr\{|x_i| \geq \sigma\epsilon\} \leq E(X^2)/(\sigma\epsilon)^2 \qquad (2.8\text{--}2)^*$$

If, as we have assumed, the mean is zero then we have

$$\Pr\{|x_i| \geq \sigma\epsilon\} \leq 1/\epsilon^2$$

In order to understand this important result we analyse the extreme of how the equality might arise. First, we dropped all the terms for which $|x_i| < \epsilon$. Second, we replaced all larger x_i by their minimum ϵ. There might have been no terms dropped in the first step, and all the terms might have been of the minimum size in the second, so no reduction occurred and the equality held; all the values were piled up at the distance ϵ from the origin, and there were no others, but this is unlikely! In this sense the Chebyshev inequality cannot be strengthened—but other forms can be found. We could, for example, use any nonnegative function in place of x_i^2 if we wish to get other inequalities.

We now turn to a closer study of the average of n independent samples of a random variable X having values x_i. Since we do not want to talk about the individual sample values we will again treat the n samples as n random variables X_i, $(i = 1, 2, \ldots, n)$. The average is defined to be

$$S(n)/n$$

where

$$S(n) = [X_1 + X_2 + \cdots + X_n] \qquad (2.8\text{--}3)$$

In the previous Section 2.7 we studied the mean and variance of $S(n)/n$ when we averaged over the whole sample space and found their corresponding values μ and σ^2/n.

Since $E\{.\}$ is a linear operator, to get the expected value of the sample average $S(n)/n$ we average over the whole sample space and get the individual

$E\{X\}$. We then apply the Chebyshev inequality (2.8-1) to the difference between $S(n)/n$ and its expected value μ,

$$\Pr\{\,|S(n)/n - \mu| \geq \epsilon\} \leq V\{S(n)/n\}/\epsilon^2 n$$

$$\leq (1/n^2)V\{S(n)\}/\epsilon^2 \qquad (2.8\text{-}5)^*$$

$$\leq V\{X\}/\epsilon^2 n$$

This is the desired result; the probability that the average of n independent samples of a random variable X differs from its expected value μ by more than a given constant ϵ is controlled by the number on the right. We see that for a random variable X by picking n large enough we can *probably* have the average of n samples close to the expected value μ of the random variable X.

There is a complement form (since the total probability must be 1), and using $k\sigma$ in place of ϵ, we get

$$\Pr\{\,|S/n - a|\} < k\sigma\} > 1 - V\{X\}/(k\sigma)^2 n \qquad (2.8\text{-}6)^*$$

that is often useful. One form says that it is unlikely that the sample average will be far out, the other that it is probable it will be close.

The results may seem to apply only to binary choices, but consider a die. The application with $p = 1/6$ applies to each face in turn, (with probability $5/6$ for all the other faces), hence each face will have the ratio of 1 to 6 for success to total trials. Similarly, any multiple choice scheme can be broken down to a sequence of two way choices, one choice for one given outcome and the other for all the rest combined, and then applied to each outcome in turn.

In this derivation we found an upper bound on the probability—but the earlier result for the binomial distribution 2.4–6 shows that the deviation of the average from the mean is actually

$$1/2\sqrt{n}$$

and suggests that this is the probable behavior in the general case; we can reasonably expect the square root of the variance to be a measure of the spread of a distribution of the average of n samples.

The idea that we may take an infinite number of samples is not practical; we often have to first decide on the number of samples n we can afford to take and then settle for what we get. For a choice of ϵ we can be, *probably*, that close *if* we accept the corresponding low level of confidence; if we want a high level of confidence we must settle for a large deviation from the expected value; there is an "exchange," or "tradeoff," of accuracy for reliability, closeness for confidence. We also see the price we must pay in repetitions of the trials for increased confidence. To get another decimal place of accuracy we must reduce the k by 10 and the formula requires an n of 100 times as large to compensate.

It is the root mean square of the variance that estimates the mean distance and the accuracy goes like $1/\sqrt{n}$; accuracy does not come cheap according to this formula!

Since the law of large numbers is widely misunderstood it is necessary to review some of the misconceptions. First, it does not say that a run of extra large (or small) values will be compensated for later on. The trials are assumed to be *independent* so that the system has no "memory"; no toss of a coin or roll of a die "knows" what has happened up to that point so it cannot "compensate." If it did then where in the whole system would the information about the past be? The law merely says that in the long run, *probably*, any particular run will be "swamped," not compensated for.

Second, the law is not a statement that the average *will be* close to its expected value—it cannot say this because there is a finite, though small, chance that an unusual run will occur (for example that 100 heads in a row will occur with an unbiased coin and the average of this run will not be close to the expected value). The law says not only that *probably* you will be close but it also implies that you must expect to be far out occasionally.

Third, we have an inequality and not an approximation. Often the Chebyshev inequality is very poor and hence makes a poor approximation when used that way. Still, as we have just seen, it can give useful results.

At first glance it is strange that from the repeated *independent* events we can (and do) find statistical regularity. It is worth serious consideration on your part to understand how this happens; how out of random independent events comes some degree of regularity.

Exercises 2.8

2.8–1 In the derivation of the Chebyshev inequality use $(x - \mu)^2$ in place of x.

2.8–2 In the derivation of the Chebyshev inequality use any $g(x) \geq 0$ you wish.

2.9 The Statistical Assignment of Probability

From the concepts of symmetry and interchangability we derived the measure of something we called *probability*. We then introduced the concept of *expected value* of a random variable which is simply the probability weighted average of the *values* of the random variable over its sample space, and it is not necessarily what you think is the "expected value."

From the model we deduced the weak law of large numbers which says that *probably* the average of many trials will be near to the expected value. From the model we found the probable frequency of occurrence. This is the justification, such as it is, for the use of the name "expected value."

Statistics, among other things, attempts to go the other way, from the observations of the frequencies of the events to deduce the probabilities of the underlying mechanisms. The basic tool is the weak law of large numbers. Some people have *postulated* the existence of these *limiting frequencies*, (see Section 8.4), but in practice, as indicated earlier, one has to settle, more or less, for a finite number of trials, and cannot hope to go to the limit. The weakness of the weak law of large numbers is both in the limit of accuracy obtainable when a realistic number of trials is assumed, and the fact that the law has the words "with probability...". It contains no certainty when you approach the matter in a scientific spirit and do not invoke impossible things like an infinite number of trials. The law tells us that in any particular case we can have no certainty as to whether we just happen to fall in the range covered by the words "with probability..." or not. The law also assumes that the variance exists and that we know it (but see Appendix 2.A), and that the average will settle down; we will later see (Chapter 8) that there are many interesting distributions for which the average does not even exist hence the weak law of large numbers does not apply!

The law of large numbers is the justification for the frequency approach as a measure of the probability of an event. When the basic symmetry of the situation is not available, as in tossing a thumb tack to see if it ends up on its side or with the point up, Figure 1.10–1, there is little else we can do but make repeated trials and note the frequencies of each from which we will deduce the probabilities to assign to the two possible outcomes of a trial. In doing so you have abandoned certainty (if you believe in the symmetry) in exchange for the possibility of doing something that may be "reasonable." The proposal that we compute the details of the model involving the tossing mechanism, elasticity and roughness of coin and surface it falls on, etc., is hardly practical (see Chapters 5 and 6).

Example 2.9–1 *Test of a Coin and Monte Carlo Methods*

The law of large numbers suggests that one way to find the probability of an event is to make many repeated trials and use the ratio of the number of successes to the total number of trials as a measure of the probability of a single event. Although we could toss a real coin many times we preferred to simulate it by using a pseudo random number generator (Section 1.9); the use of a computer tends to reduce the errors in an experiment using many trials (which after a while tends to become boring and induces carelessness). This simulation process assumes that we have a model that captures the essence of the situation.

We used a pseudo random number generator and assigned those numbers greater than 1/2 as H and the others as T. The results of 2000 trials on two different computers using different pseudo random number generators gave Table 2.9-1.

TABLE 2.9–1

			Deviations from expected	
n	run #1	run #2	run#1	run #2
200	108	101	8	1
400	213	219	13	19
600	327	323	27	23
800	434	419	34	19
1000	530	513	30	13
1200	619	614	19	14
1400	709	715	9	15
1600	799	814	−1	14
1800	897	909	−3	9
2000	1008	1011	8	11

We used the deviation from $p = 1/2$ as the measure of error. Does what we see seem to be reasonable? The weak law of large numbers gave one inequality. We have seen that for binomial trials it is appropriate and will later show in Chapter 9 that it is reasonable, at times, to take the square root of the variance as a measure of the deviation.

From (2.8-3) we have (where S is the sum)

$$\Pr\{|S/n - 1/2| \geq \epsilon\} \leq V\{X\}/\epsilon^2 n$$

The variance $V\{X\}$ is given by

$$V\{X\} = (0 - 1/2)^2 + (1 - 1/2)^2 = 1/2$$

Thus we have (multiply by n inside the probability)

$$\Pr\{|S - n/2| \geq n\epsilon\} \leq 1/2n\epsilon^2$$

If we now pick $\epsilon = 25/n$ then we have

$$\Pr\left\{\,|\,S - n/2\,| \geq 25\right\} \leq n/1250$$

Thus the results seem reasonable. One sees the regularity of the deviations in each case because we are computing the cumulative sums. The law indicates that one should expect deviations frequently. The distributions are cummulative and any early deviation tends to persist in the following lines of the table until it is "swamped" by the rest.

As we will see, there are other things we can deduce from the model which will encourage us to adopt the reverse reasoning, to go from the frequency data to the probabilities of the model. But when everything is faced, the path from the data back to the model is not as secure as we wish it were; but what else can we do? To do nothing in the face of uncertainty can be more dangerous than acting with reasonable prudence. Hence the important, highly developed field called *statistics* is worth cultivating, but it lies outside the province of this book.

Example 2.9–2   *Chevalier de Mere Problem*

It is said that the Chevalier de Mere observed that the probability of getting at least one 6 in four tosses of a die is greater than the probability of getting at least one double 6 is 24 throws of two dice.

The probability of not getting a 6 in four throws is

$$(1 - 1/6)^4$$

hence the probability of getting at least one 6 is

$$1 - (1 - 1/6)^4 = 1 - (5/6)^4 = 0.51775\ldots$$

The probability of getting at least one double 6 in 24 throws is similarly

$$1 - (1 - 1/36)^{24} = 1 - (35/36)^{24} = 0.49140\ldots$$

We may well wonder how many trials de Mere must have made (if he did make the experimental observation at all) to find this delicate difference. This is not a book on statistics, yet the topic is of sufficient importance that we will make a crude estimate.

As a first approach we note that each binomial distribution about its mean has a corresponding variance, and that it is reasonable to ask that the two distributions be separated by the sum of the variances—else the two distributions would appear as one single one, Figure 2.9-1. The difference of the means is

$$0.51775 - 0.49140 = 0.02635$$

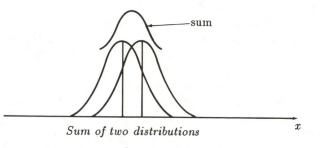

Sum of two distributions

x

FIGURE 2.9–1

and we allot one half of this amount, namely 0.0132, to the variability of each then we get from (3.7–5) and (3.7–6)

$$\mu = np \qquad \sigma^2 = npq$$

The product pq is in both cases about 1/4. Hence we estimate the number of trials as

$$n = pq/(0.0132)^2 \sim 1500$$

If this is used we get a total of 6000 rolls for 4 tosses to get a 6. For the 24 throws of two dice we have 48 rolls each, so we get 72,000 or a total of over 78,000 rolls of a die. One can only wonder at the ability to keep such a record intuitively in one's head. Of course we could better allocate the total variance to minimize the total number of rolls, but this does not reduce things enough to make the difference being examined easy to recognize.

2.10   The Representation of Information

Now that we have a connection between the symmetry definition of probability and the commonly assumed frequency definition, we can see many situations in both forms (provided the mean exists). An important field of application is that of representing information. Due to technical reasons we currently use the binary form of representation, two marks, either a 0 and a 1, or as in the game of tic-tac-toe a circle and a cross. The amount of information is often called a *bit* (an abreviation of binary digit).

The ASCII code is an example where each symbol in the alphabet is represented by a string of 7 bits. Thus there are $2^7 = 128$ possible symbols, and these include in this code the Roman alphabet, both lower and upper case, the decimal digits, a lot of punctuation, and a large number of special codes.

It is customary in sending the ASCII code symbols to append an eighth bit at the end, and choosing this digit to be such that the total number of

1s in the symbol is even (or odd). This permits error detection since a single change in a bit will be recognized at the receiving end. In the simple theory it is customary to assume "white noise" meaning simply the Bernoulli model of constant probability and independence of errors in various positions. This use of an even (odd) number of 1s in the message is called a *parity check*.

Similar error detecting codes are widely used in our society and the field is highly developed; hence we can mention only one simple case, the two-out-of-five code that was widely used in the telephone system. It is based on the idea that since $C(5,2) = 10$, the decimal digits can be represented by exactly two 1s in the five positions (of the block code in which each symbol is of length 5 bits).

We can go farther with parity checks and put in enough different parity checks to locate the position of the error, and hence be able to correct it at the receiving end by complementing the received bit in that position. We illustrate it by the simple case of four message positions and three parity checks. The checks are:

check number	positions of the parity check
#1	1, 3, 5, 7
#2	2, 3, 6, 7
#3	4, 5, 6, 7

and for convenience we put the parity check bits in the positions 1, 2 and 4. To see how it works suppose we have the four bit message 1011 to send. It must go into positions 3, 5, 6, 7. We have the following table where we suppose that there is an error in position 6.

positions	1	2	3	4	5	6	7
message	–	–	1	–	0	1	1
encoded message	0	1	1	0	0	1	1
error						x	
received message	0	1	1	0	0	0	1

At the receiving end you apply the parity checks writing (from right to left) a 0 if the check is correct and a 1 if it is not. You get for the #1 check 0, for #2 a 1, and for #3 a 1. From right to left this is the binary number

110 which is decimal 6

so you change the digit in position 6, strip off the parity checks and you have the message.

This seems very inefficient for this short code, but a moment's reflection will show you that for k parity checks you get a total message length of $2^k - 1$ (of which k positions are used for the parity checks. Thus at $k = 10$, there are 1023 positions of which 1013 are useful message positions.

Exercises 2.10

2.10–1 In the code with $n - 1$ message positions and one check position what is the probability of no error? Ans. $(1 - p)^n$

2.10–2 In example 1 if $n = 1/p$ (large) what is the probability of no error? Ans. $1/e$.

2.10–3 In the 2-out-of-5 code what is the probability of an error? Ans. $C(5,1)pq^4$

2.10–4 In the 2-out-of-5 code what is the probabilty that two errors will not be detected?

2.10–5 In an error correcting code with $k = 10$, what is the probability of an undetected error? Note that all odd numbers of errors will be detected.

2.10–6 In the error correcting code if an additional parity check over the whhole message is added, show that double errors will be detected but will not be correctable.

2.10–7 In the code of Exercise 2.10–6 estimate the probability of an undetected error if p is small.

2.10–8 Discuss the wisdom of using the extra parity check (see Exercise 2.10–6) to detect double errors.

2.11 Summary

In Chapter 1 we introduced a simple model for the probability of an event based on symmetry, as well as several related ideas such as the sample space and randomness.

In this chapter we have developed the mathematical concepts needed to proceed, such as permutations and combinations, the binomial distribution, random variables, the mean and variance, generating functions and convolutions, and the weak law of large numbers which is central to the connection between the probability of a simple event and the frequency of occurence in a long run of identical and independent trials. We saw that the two concepts of probability, one based on symmetry and the other on the frequency of occurence, are *not* equivalent, that for the law to apply the variance must exist (but see Appendix 2.A).

In the next Chapter we will turn to developing five mathematical methods for solving problems. By examining the methods we will try to avoid Peirce's condemnation of the reliability of solving probability problems; we will see the methods more clearly and become more familiar with their use and misuse.

Appendix 2.A   *Derivation of the Weak Law of Large Numbers*

The weak law of large numbers was made to depend on the existence of the variance; it can be freed of this constraint as follows - though it still depends on the existence of the mean.

The approach is the classical one of the calculus; treat the infinite range as a finite range, compute the result, and then examine the result as the range is allowed to approach infinity.

The method is sometimes called *the method of truncated variables* and is quite useful in general. We assume we are looking at the average of n obervations of a random variable X. The expected value of X can be taken, for convenience, to be 0 since we could replace the random variable X by an new random variable $X - \mu$, where μ is the mean. We regard the average (as before) as the average of n random variables X_i, $(1 = 1, 2, \ldots, n)$. To cope with the limit these variables X_i are then truncated

$$\begin{cases} U_i = X_i \\ U_i = 0 \end{cases} \qquad \begin{cases} V_i = 0 \\ V_i = X_i \end{cases} \qquad \begin{array}{l} \text{if } |X_i| \le kn \\ \text{if } |X_i| > kn \end{array}$$

where k is some (small) positive constant to be determined later. Thus we have for all i

$$X_i = U_i + V_i$$

To prove the weak law of large numbers we need to show that for any $e > 0$ both

$$P\{|U_1 + U_2 + \cdots + U_n| > \epsilon n/2\} \to 0$$

$$P\{|V_1 + V_2 + \cdots + V_n| > \epsilon n/2\} \to 0$$

Notice that we are considering only finite n; for any given ϵ there is a corresponding n. Let the possible values of X_i be $x_1, x_2, \ldots$ and their corresponding probabilities be $p(x_j)$. Then the expected value of $|X_i|$ is

$$\sum_j |x_j| p(x_j) = a$$

The bound on the U_i means that the x_j are also bounded, hence since the U_i are bounded by kn we can take one factor out and have

$$E\{U_i^2\} \le kn \sum_j |x_j| p(x_j) \le akn$$

The variables U_i all have the same probability distribution and are independent, hence

$$V\{U_1 + \cdots + U_n\} = n\, E\{U_i^2\} \le akn^2$$

But from the definition of the truncated variables as $n \to \infty$

$$E\{U_i\} \to E\{X_i\} = 0$$

Hence for sufficiently large n (since the X_i are independent the U_i are independent)

$$E\{(U_1 + \cdots + U_n)^2\} \leq akn^2$$

We now apply the Chebyshev inequality to get

$$P\{|U_1 + \ldots + U_n| > \epsilon n/2\} \geq 8ak/\epsilon^2$$

and by choosing the k small enough our first requirement on the truncated variables is realized.

We now turn to the V_i. We have immediately

$$P\{V_1 + \cdots + V_n \neq 0\} \leq n\, P\{V_i \neq 0\}$$

For arbitrary $k > 0$ we have

$$P\{V_i \neq 0\} = P\{|x_i| > kn\} = \sum_{|x_i| > kn} p(x_i)$$

$$\leq (1/kn) \sum_{|x_i| > kn} |x_i|\, p(x_i)$$

and this sum approaches 0 as n becomes infinite. Both terms approach zero so that the sum may be made as small as desired.

Appendix 2.B   *Useful Binomial Identities*

Binomial identities play a large role in probability and other fields of mathematics, hence they are worth some attention. Since there are potentially an infinite number of identities among the binomial coefficients we will give only a few of the more useful ones.

From the definition

$$C(n, k) = \frac{n!}{k!(n-k)!} = \frac{n!(n-k-1)}{k(k-1)!(n-k+1)!}$$

$$= \frac{(n-k+1)}{k} C(n, k-1) \tag{2.B-1}$$

$$C(n, 0) = 1$$

we can compute the binomial coefficients recursively (as was used in (2.4–3)).

We often need the binomial coefficients for negative order. In the definition

$$C(n, k) = \frac{n(n - 1)(n - 2)\ldots(n - k + 1)}{k!}$$

we simply replace n by $-n$ and get

$$C(-n, k) = \frac{(-1)^k(n + k - 1)(n + k - 2)\ldots(n)}{k!} = (-1)^k C(n + k - 1, k)$$

$$(2.\text{B}-2)$$

If we multiply two binomial generating functions of different orders together we get

$$(1 + t)^{m+n} = (1 + t)^m(1 + t)^n$$

$$\sum_k C(m + n, k)t^k = \sum_i C(m, i)t^i \sum_j C(n, j)t^j \qquad (2.\text{B}-3)$$

Pick out the kth power of t

$$C(m + n, k) = \sum_i C(m, i)C(n, k - i)$$

For the special case $m = n = k$ we get

$$C(2n, n) = \sum_i C(n, i)C(n, n - i)$$

$$(2.\text{B}-4)$$

$$= \sum_i C^2(n, i)$$

In words, the sum of the squares of the binomial coefficients of order n is exactly the mid-coefficient of the expansion of order $2n$.

Similarly identities like

$$1 + C(m + 1, m) + C(m + 2, m) + \cdots + C(m + n, m) = C(m + n,, m + 1)$$

are easily proved.

3

Methods for Solving Problems

"You know my methods, Watson."
Sherlock Holmes

3.1 The Five Methods

It is widely believed that the only way to learn to do probability problems is by doing them. We have already used the method of "compute the complement" and the method of simulation. In this Chapter we will give and illustrate five widely used general methods for solving problems. They are:

1. *Total sample space.* First write out the total sample space.

(A) If the elementary events in the sample space are equally likely then the ratio of the number of successes to the total number of points in the sample space is the probability of a success.

(B) If the elementary events in the sample space have different probabilities then the probability of a success is the sum of all the probabilities of the elementary events that are successes.

We see that (A) is a very common special case of (B).

2. *Enumerate.* Enumerate (count) somehow only the equally likely successes and divide by the size of the total sample space to get the probability of a success (or add the probabilities of the successes).

3. *Historical.* Follow the history of the independent choices and take the product of the probabilities of the individual independent steps.

4. *Recursive.* Cast the problem in some recursive form and solve it by induction or recursion. This is often easiest done using *state diagrams* which

are increasingly used in many parts of science. We shall develop and elaborate
this method in Chapter 4.

   5. *Random Variables.* In Section 2.5 we carefully introduced the idea of
a random variable which conventionally written as X. It is used to indicate
the value associated with the outcome of a general trial X (same label) whose
possible outcomes are the points in the sample space. The outcomes are the x_k
with probabilities $p(k)$ and often having associated values that are simply k.
Using the notation of Sections 2.5 and 2.6 we have for the expected value and
variance (for $x_k = k$), $(0 \leq k \leq K)$

$$E\{X\} = \sum_{k=0}^{K} kp(k) = \mu$$

$$V\{X\} = \sum_{k=0}^{K}(k - \mu)^2 p(k) = E\{X^2\} - [E\{X\}]^2 = \sigma^2$$

   It is often easy to solve fairly complex problems using random variables
(and a different frame of mind from the usual approach to probability problems).

   We will first use a simple "occupancy" problem to illustrate the methods
by solving the same problem by all five methods. Occupancy problems are
the classic way of posing many probability problems—the standard form is:
"In how many ways can a certain number of balls be put in a given number
of places subject to the following restrictions?" The occupancy solution is
readily converted to probabilities.

Example 3.1–1   The Elevator Problem (Five Ways)

The problem we will solve in this section is: suppose an elevator car has 3
occupants, and there are 3 possible floors on which they can get out. What
is the probability that exactly 1 person will get out on each floor?

   Since nothing else was said, we *assume* that each person acts independently of the others, and that each person has an equally likely chance of
getting off at each floor. Without some such assumptions there can be no
definite problem to solve. Of course different assumptions generally lead to
different results.

   Method 1. **Total Sample Space.** We first create the total sample
space. Let the people be labeled a, b and c, and list in some orderly fashion
all possible choices of the exit floor for each person. One such ordering is
shown in Table 3.1–1.

TABLE 3.1–1

Total Sample Space

The total sample space showing floor number (1, 2, or 3)

a	b	c		a	b	c		a	b	c
1	1	1		2	1	1		3	1	1
1	1	2		2	1	2		3	1	2*
1	1	3		2	1	3*		3	1	3
1	2	1		2	2	1		3	2	1*
1	2	2		2	2	2		3	2	2
1	2	3*		2	2	3		3	2	3
1	3	1		2	3	1*		3	3	1
1	3	2*		2	3	2		3	3	2
1	3	3		2	3	3		3	3	3

The $3^3 = 27$ different cases in the product sample space are all equally likely since we assumed both the independence of the people and the uniform probability choice for each floor. The six starred entries meet the criterion that all three floors are distinct, hence the probability we seek is

$$6/27 = 2/9$$

Method 2. **Enumeration of Successes.** In this method we write down only the 6 successes, $(1, 2, 3)$, $(1, 3, 2)$, $(2, 1, 3)$, $(2, 3, 1)$, $(3, 1, 2)$, $(3, 2, 1)$. We could also note that the number of ways of getting a success is exactly $P(3, 3) = 6$. The total sample space is $3 \times 3 \times 3 = 27$ (each person may choose any floor). Hence the probability sought is

$$P(3, 3)/3^3 = 6/27 = 2/9$$

Method 3. **Historical.** In this method we trace the history of the elevator. For a success at the first floor only one person can get off. For the second stop there is again a probability of success. For the third similarly—hence the answer is the product of these three probabilities.

At the first floor if only one person gets off (and two do not) then, since there are 3 ways of choosing the person who gets off, we have $3[(1/3)(2/3)^2] = 4/9$. Using the earlier notation (2.4–2) we get the same result

$$b(1; 3, 1/3) = C(3, 1)(1/3)(2/3)^2 = 12/27 = 4/9$$

If we have a success at the first floor then at the second floor we have only two people left, and hence we have

$$2(1/2)(1/2) = 1/2$$

or in the earlier notation

$$b(1; 2, 1/2) = C(2, 1)(1/2)(1/2) = 2/4 = 1/2$$

At the third floor the probability of success (assuming that the first two are successes) is exactly 1, hence for all three floors to be successes

$$(4/9)(1/2)(1) = 2/9$$

as before.

Method 4. **Recursive.** To make the recursive method easy to understand we generalize the elevator problem to the case of n stops and n people. Let $P(n)$ be the probability that exactly one person gets off at each of the n stops. Examining the first stop we have for a success on that floor that exactly one of the n persons gets off (with probability $1/n$) and the other $n-1$ do not get off (with probability $[(n-1)/n]^n - 1$), then the recurrence formula is, since we have $n-1$ persons left and $n-1$ floors,

$$P(n) = (n)(1/n)[(n-1)/n]^{n-1}P(n-1) = [(n-1)/n]^{n-1}P(n-1)$$

$$P(1) = 1$$

which is a simple recursion relation. From (2.4–2) we could have written the coeficient as $b(1; n, 1/n)$ as we did the earlier ones.

We now unravel the recursion.

$$P(n) = [(n-1)/n]^{n-1}P(n-1)$$

$$= [(n-1)/n]^{n-1}[(n-2)/(n-1)]^{n-2} \ldots [1/2][1]$$

$$= (n-1)[1/n]^{n-1}(n-2)(n-3) \ldots 1$$

$$= (n-1)!/n^{n-1} = n!/n^n$$

For $n = 1$ this gives 1 as it should; for $n = 2$ we get $1/2$ which is clearly correct as can be seen from the sample space of four points; and for $n = 3$ we get $6/27 = 2/9$ as before. In some respects the general case is easier to solve than is the particular case of $n = 3$.

The general solution usually provides simple checks along the way to increase our confidence in the answer. If we use Stirling's formula for the factorial (see Appendix 1.A) we get

$$P(n) \sim e^{-n}\sqrt{2\pi n}$$

and even at $n = 3$ the approximation is close. The following Table 3.1–2 indicates both: (1) how rapidly the approximation approaches the true value, and (2) how rapidly the probability approaches 0 as the number of floors increases.

TABLE 3.1–2

Elevator Problem

n	estimated	true	true − estimate
1	0.9221...	1.0000	0.0779...
2	0.4798...	0.5000... = 1/2	0.0202...
3	0.2161...	0.2222... = 2/9	0.0061...
4	0.0918...	0.0937... = 3/32	0.0019...
5	0.0377...	0.0384... = 24/625	0.0007...
6	0.0152...	0.0154... = 5/324	0.0002...
7	0.0060...	0.0061... = $6!/7^6$	0.0001...
8	0.0024...	0.0024... = $7!/8^7$	
9	0.0009...	0.0009... = $8!/9^8$	
10	0.0004...	0.0004... = $9!/10^9$	

This table gives us some feeling for the situation as a function of n, the number of floors.

The recursive method soon becomes the method of difference equations and leads to state diagrams, hence its importance is greater than it may seem now.

Method 5. **Random Variables.** We begin by assigning the random variable X to be 1 if at each floor only one person gets off and 0 otherwise. Thus the expected value of X will be the probability we want. Next we ask, "How can we break down this random variable into smaller parts?" Let the random variable $Y_i, (i = 1, 2, 3)$, be 1 if the ith floor is not used and 0 if it is. Then we have the key equation

$$X = (1 - Y_1)(1 - Y_2)(1 - Y_3)$$

We now have

$$P = E\{X\}$$
$$= E\{1 - [Y_1 + Y_2 + Y_3] + [Y_1 Y_2 + Y_2 Y_3 + Y_3 Y_1] - Y_1 Y_2 Y_3\}$$
$$= 1 - [p(1) + p(2) + p(3)] + [p(1, 2) + p(2, 3) + p(3, 1)] - p(1, 2, 3)$$

where

$p(i)$ = probability that the ith floor is not used

$p(i, j)$ = probability that both the ith and jth floors are not used

$p(i, j, k)$ = probability that none of the floors is used.

The last sum is over distinct i, j and k, and is, of course, 0 since at least one floor must be used. We have

$$p(i) = (2/3)^3$$

$$p(i, j) = (1/3)^3$$

$$p(i, j, k) = 0$$

Putting these in the formula we get

$$P = 1 - 3(2/3)^3 + 3(1/3)^3 - 0 = 1 - 8/9 + 1/9 = 2/9$$

which is again the same answer.

The method of random variables seems to be a bit laborious in this simple example, but in section 3.6 we will give other examples which begin to show the power of the method. It should be clear that it requires a different way (art) of thinking, and that you need to learn how to pick wisely the random variables you define.

We now compare the five methods. When you can do the complete listing of the sample space it gives you a feeling of safety. When you try to enumerate the successes you are already a bit worried that you may have made an error. The historical approach also has some worries that you may have made an error but it seems to be relatively safe when you can use it. The recursive method, especially when you can generalize the problem, gives supporting checks of various cases and concentrates almost all the attention at two places, the setting up of the recurrence equation and the initial conditions. The random variable method is most effective in complex cases and has an elegance of method that makes it attractive once it is mastered. It seems to be relatively safe—when it works!

We next turn to the development of the five methods, and along the way we introduce a number of other useful ideas that arise in probability.

3.2 Total Sample Space and Fair Games

The method of laying out the total sample space and counting the successes is very simple to understand, hence in illustrating it we will introduce a new idea, that of a *fair game*.

A "fair game" is usually defined to be one in which your expectation of winning (probability of winning times the amount you can win) is equal to your expectation of losing (probability of losing times the amount you can lose). Thus for a fair game your total expectation is exactly 0. For example, betting a unit of money on the toss of a "fair" coin, $\Pr\{H\} = 1/2$, and getting two units of money if you win (you win one unit) and none if you lose (you do not get your money back) is a fair game. Your total expectation is

$$(1/2)(1) + (1/2)(-1) = 0$$

and it is by definition a *fair game*. There are in the literature much more elaborate and artificial definitions designed mainly to escape later paradoxes; we will stick to the common sense definition of a fair game.

Example 3.2–1   *Matching Pennies (Coins)*

A common game to consume time is "matching pennies" in which two players A and B each toss a coin (at random and independently). If the two coins show the same face then A wins (gets both coins), and if they are opposite then B wins. Supposing that the coins are each "well balanced" (meaning that the probability of a head is exactly $1/2$) then the complete sample space is:

$$HH \qquad HT \qquad TH \qquad TT$$

each with probability of $1/4$. A wins on HH and TT so A has a probability of $1/2$; so has B. Thus it is an unbiased trial and with the equal payoffs it is a fair trial.

We will use random variables to handle the repeated (independent) trials. Let X_i be the random variable that on the ith toss A wins. It has the values

$$X_i = \begin{cases} 1 & \text{win} \\ -1 & \text{lose} \end{cases}$$

After n tosses we have the random variable S_n for the sum

$$S_n = X_i + X_2 + \cdots + X_n$$

as the value of the game to A. It is easy to see from the total sample space that for each i the $E\{X_i\} = 0$, and $V\{X_i\} = 1$; each independent trial is fair and you either win or lose. Now for S_n we have

$$E\{S_n\} = 0$$
$$V\{S_n\} = n$$

Since the expected value is 0 it is a fair game. Repeating a fair trial many times is clearly a fair game.

But by the law of large numbers (Section 2.8) it is the average, S_n/n, that probably approaches 0, not S_n. Indeed, we have already remarked that it is reasonable to estimate the deviation from the expected sum by an amount proportional to the square root of the variance (since the variance is the mean square of the deviations from the expected value). Hence we more or less "expect" that one player will have gained, and the other player lost, an amount about $\sqrt{n}$, though the expectation is 0. Yes, it is a fair game, but we "expect" that after n games one of the players will have lost approximately the amount $\sqrt{n}$ of money.

Example 3.2–2   *Biased Coins*

Suppose in Example 3.2–1 that the two coins are not exactly evenly balanced; rather that for the first coin the probability of a head is

$$1/2 + e_1 \qquad |e_1| < 1/2$$

and for the second coin it is

$$1/2 + e_2 \qquad |e_2| < 1/2$$

The probability that A wins—both coins are either heads or both tails—is now

$$(1/2 + e_1)(1/2 + e_2) + (1/2 - e_1)(1/2 - e_2) =$$

$$= 1/2 + 2e_1e_2$$

and of course for B it is the complement probability $1/2 - 2e_1e_2$.

In this case we see that the fairness of this game is not changed much by small biases in the coins—that the effect is of the order of the product of the two biases e_i, which should be quite small in practice. Such a result is said to be *robust*—that it is relatively *insensitive to small changes* in the assumed probabilities. Such situations are of great importance since in practice we seldom know that the probabilites in the real world are exactly what we assume they are in the probability model we use. See Section 3.9.

Example 3.2–3   *Raffles and Lotteries*

We consider first the simple raffle. In a raffle there may be several prizes, but for simplicity we will suppose that there is only one prize of value A (in some units of money). Let there be n tickets sold, each of unit value. What is your expected gain (assuming that each ticket has an equally likely chance of being drawn)? The probability, per ticket, is $1/n$, so your expected gain G is

$$G = (A/n) \text{ (number of tickets you hold)} - \text{(cost of your tickets)}$$

If the total number of tickets sold (yours plus the others) were exactly $n = A$, and if you hold k tickets, then your expected gain would be

$$G = (k/n)[A] - k = k - k = 0$$

and it is "fair." If more tickets are sold then it is "unfair to you" and if less then it is an "advantage to you."

If you enter many similar "fair" raffles then according to the weak law of large numbers you will come close to balancing your losses and winnings, but the deviation between them will tend to go like $\sqrt{n}$.

To review, in a "fair game" your expectation of winning (per independent trial) is the same as the price you pay—the net expectation is zero. Otherwise it is "unfair" one way or the other. You may, of course, engage in a game that is unfair to you for purposes of amusement, or to "kill time," and there are other reasons for buying raffle tickets—excitement, contributions to charity, easing your concience, obliging friends, etc. beyond just entering as a means of making money. As a general rule, there are far more tickets sold in a raffle than the value of the prize (prizes), and the raffle/lottery is thus unfair to you, but occasionally you see the opposite—due to charitable donations of prizes and/or the failure to sell enough tickets (possibly due to a stretch of bad weather) there may be a favorable raffle for you to enter.

In some kinds of lotteries it may happen that for several times no one has won the prize and it has accumulated. Thus it is conceivable that a lottery may be advantageous to you—but it is unlikely!

These are simple examples of looking at the total sample space; it is the number of tickets sold. Assuming the fair drawing of the tickets your chance of winning is the number you hold divided by the total sold (including yours). Your net expectation is this probability multiplied by the value of the prize minus your cost of buying tickets. This raises the following simple problem.

Example 3.2–4   *How Many Tickets to Buy in a Raffle*

Assuming: (1) the value of the prize is A, (2) that you know the number n of tickets sold (of unit cost), and (3) that you can buy as many as you wish at the last moment (and no one else can follow after you), then how many tickets should you buy?

Let the number to buy be k. After you buy k tickets then there are $n+k$ tickets sold, and your probability of winning is $k/(n+k)$. Your expectation of winning is $Ak/(n+k)$, and your *gain* is

$$G(k) = Ak/(n+k) - k$$

We want to maximize $G(k)$. We first *assume* that $G(k)$ is a continuous function of k so we can apply the calculus (though we know it is discrete because we must buy a whole ticket at a time—still it won't be far off, we hope). We can now scale the problem and isolate the size A of the prize. Using the value A of the prize as a measure, we introduce the *relative* number of tickets

$$n/A = u, \quad \text{or} \quad n = uA$$

$$k/A = v \quad \text{or} \quad k = vA$$

as new (relative) variables. We have, therefore, the corresponding continuous function

$$G(vA) = \frac{AvA}{uA + vA} - vA = A\left[\frac{v}{u+v} - v\right]$$

To find the maximum we set the derivative with respect to v (the variable of the problem; u is known) equal to zero.

$$AG'(vA) = A\left[\frac{(u+v) - v}{(u+v)^2} - 1\right] = 0$$

We get, when we multiply through by the denominator,

$$(u+v)^2 = u$$

from which we get

$$v = \sqrt{u} - u$$

Going back to the original variables n and k we have the number of tickets to buy is

$$k = A[\sqrt{u} - u] = A[\sqrt{n/A} - n/A] = \sqrt{nA} - n$$

Since we are imagining that $u < 1$ (or else we are not interested in the raffle) we have $v > 0$. At this optimum the value (gain) of the raffle to us is, of course,

$$G(v) = A \left[\frac{\sqrt{u} - u}{u + \sqrt{u} - u} - \sqrt{u} + u \right]$$

$$= A[1 - \sqrt{u} - \sqrt{u} + u] = A(1 - \sqrt{u})^2$$

$$= [\sqrt{A} - \sqrt{n}]^2$$

We see that when u is near to 1, that is n is near A (the value of the tickets sold is almost the value of the prize) then the value of the raffle to us is very small indeed—especially if you look at the risk you take in making the investment.

For the original discrete variable problem in integers these answers are not exactly correct since we replaced the discrete problem by a continuous one and solved the new problem. For example, in the extreme case of $n = 0$ (no tickets sold) you should buy 1 ticket, but the formula indicates no tickets. When $u = 1$, namely $n = A$, there is no sense in buying a ticket since then your expectation for the first ticket is $A/(n+1) = n/(n+1)$ and is less than what you will pay for the ticket. Indeed, at $n = A - 1$ it is doubtful that it is worth the effort, though the game is then fair.

Exercises 3.2

3.2–1. In Example 3.2–2 suppose that the two biases are exactly opposite, $c_1 = -c_2$. Discuss why the answer is what it is.

3.2–2 Consider Example 3.2–2 where only one coin is biased. What does this mean? Ans. The bias has no effect!

3.2–3 Discuss the case when there are several prizes in a raffle of varying values.

3.2–4 What is the probability of 3 or 1 heads on the toss of three biased coins? Discuss the solution. Ans. $1/2 + 4c_1 c_2 c_3$, and at least one fair coin makes the game fair.

3.2–5 Solve Example 3.2–5 in integers by using $G(k+1) - G(k)$.

3.2–6 In the gold-silver coin problem (Example 1.9–1) suppose that there are three kinds of coins, gold, silver, and copper, with two to a drawer, and 6 drawers. Show that the probability of a gold coin on the second draw is $1/2$.

3.2–7 In Example 1.9–1 suppose there are 4 drawers with GGG, GGS, GSS, and SSS. Show that the answer is still $2/3$.

3.2–8 If A has 3 tickets in a raffle with 9 tickets sold, and B holds 1 in a raffle of 3, find the ratio of their probabilites of having a winning ticket.

3.2–9 If there are three prizes in a raffle and A holds 3 of exactly 9 tickets sold, what is the probability that A will not get any prize? Ans. $16/21$.

3.2–10 On the roll of a die you win the amount of the face. If the game is to be fair what should you pay for a trial? Ans. 7/2.

3.2–11 You roll two dice and get paid the sum of the faces; what should you pay if it is to be fair?

3.2–12 You draw a card at random from a deck. If you get paid 1 unit for each face card and 2 units for an ace (and nothing otherwise) what is a fair price for the game? Ans. 5/13.

3.2–13 You toss 6 coins and win when you get exactly three heads. What is a fair price?

3.2–14 You roll a die and toss a coin. You get the square of the face of the die if and only if you got a head on the coin. What is a fair price?

3.2–15 You toss a coin and roll a die and are paid only if it is a H and a 4 that shows. What is a fair price? Ans. 1/12.

3.2–16 You roll three dice and are paid only if all three faces are even. What is a fair price?

3.3 Enumeration

In this method we do not try to lay out the whole sample space of equally likely cases; rather we compute its size and find only the successes.

Example 3.3–1   *The Game of 6 dice*

There is a game in which you roll 6 dice and you are paid according to the following rule:

if 1 face shows 6,	you win 1 unit
if 2 faces show 6,	you win 1 1/2 units
if 3 faces show 6,	you win 2 units
if 4 faces show 6,	you win 2 1/2 units
if 5 faces show 6,	you win 3 units
if 6 faces show 6,	you win 3 1/2 units

The argument (given by the gambler offering you the chance) for why you should play is that you expect that you will get at least one 6 on each roll of 6 dice, and all the higher ones are gains, so that if you pay 1 unit to play then it is favorable to you. But is it? (It is fairly safe to assume that any widely played game is favorable to the person running it and not to the player.)

Using the earlier notation (2.4–2) we have the probability for exactly k faces is

$$b(k; 6, 1/6)$$

From the above table the formula for the payoff for tossing k 6s

$$1/2 + k/2 \qquad (k = 1, 2, \ldots 6)$$

The expectation of the sum of the payoffs is

$$\sum_{k=1}^{6} (1/2 + k/2) b(k; 6, 1/6)$$

$$= (1/2) \sum_k b(k; 6, 1/6) + (1/2) \sum_k k b(k; 6, 1/6)$$

The first summation is (since the sum of all the probabilities from 0 to 6 is 1)

$$1/2[1 - \text{the zeroth term}] = (1/2)[1 - b(0; 6, 1/6)]$$

$$= 1/2 - (1/2)(5/6)^6$$

$$= 1/2 - 0.16745\ldots$$

The second summation can be found from the generating function of the $b(k; 6, 1/6)$, namely (2.4–5)

$$(q + pt)^6 = \sum_{k=0}^{6} b(k; 6, p) t^k$$

where $p = 1/6$ and $q = 5/6$. Differentiating with respect to t and setting $t = 1$ we get (the expected value)

$$6p = 1 = \sum_{k=1}^{6} k b(k; 6, 1/6)$$

hence we have for the payoff

$$1/2[1 - b(0; 6, 1/6)] + 1/2 = 1 - 0.16745\ldots = 0.83255\ldots$$

and the game is decidedly unfair to the player!

When you try to think about how this game can be unfair it may be confusing because the derivation is so slick. Hence to develop your intuition we write out some of the details in Table 3.3–1.

TABLE 3.3-1

The Six Dice Game

No. of 6s	probability	
0	$1(5/6)^6$	$= .334\ 898\ 0$
1	$6(5/6)^5(1/6)$	$= .401\ 877\ 6$
2	$15(5/6)^4(1/6)^2$	$= .200\ 938\ 8$
3	$20(5/6)^3(1/6)^3$	$= .053\ 583\ 7$
4	$15(5/6)^2(1/6)^4$	$= .008\ 037\ 6$
5	$6(5/6)(1/6)^5$	$= .000\ 643\ 0$
6	$1(1/6)^6$	$= .000\ 021\ 4$

We see that only about 40% of the time do we get exactly one 6 and come out even, while about 33% of the time we get no 6s at all! The gains for the multiple 6s do not equal the losses.

Perhaps why it is unfair is still not as clear as you could wish if you are to develop your intuition, so you do the standard scientific step and reduce the problem to a simpler one that appears to contain the difficulty. Consider tossing two fair coins, with payoffs of 1 for a single head and 3/2 for a double head. Now you have the simple situation

TABLE 3.3–2

The Two Coin Game

Outcome	probability	payoff
TT	1/4	0
TH	1/4	1
HT	1/4	1
HH	1/4	3/2

The total payoff (each multiplied by its probability of 1/4) is

$$(1/4)[0 + 1 + 1 + 3/2) = (1/4)(7/2) = 7/8$$

and you see that it is again unfair. But this time it is so simple that you see clearly how the result arises—it is the failure to payoff properly on the two heads, which should be one unit per head tossed. With the 6 dice the situation is sufficiently complicated that you might get lost in the details and not see that it is the failure to payoff 1 unit for *each* 6 that causes the loss. The smaller payments, by advancing by 1/2 instead of 1 each time another 6 arises, is where the loss occurs. Since you expect one 6 on the average, you lose on the multiple cases where you are not paid 1 for each 6.

The problem of finding the successful cases among the mass of all possible cases closely resembles the problem of finding a good strategy in many AI (artificial intelligence) problems. The main difference is that in AI it is often necessary to find only one good solution, while we require finding *all* solutions that meet a criterion. Still, many of their methods may be adapted to our needs.

An example of an AI technique is "backtracking" which systematically eliminates many branches where it is not worth searching. We will illustrate the method by a simple problem.

Example 3.3–2   *The Sum of Three Dice*

What is the probability that the sum of the faces on three dice thrown at random will total not more than 7?

We begin by listing the canonical (increasing) representations in a systematic fashion. We start with the first two dice having faces 1,1, and then let the third one be in turn 1, 2, 3, 4, 5. We cannot allow 6 as the sum would be too large. We now "backtrack" and advance the immediately previous face to 2, thus we have 1,2 and start with the third die running through 2, 3, 4, and that is all. And so we go, and when one face cannot be advanced further we advance the immediately previous die face one more. In this way we generate the table of canonical representations

(1,1,1)	(1,1,2)	(1,1,3)	(1,1,4)	(1,1,5)	13
(1,2,2)	(1,2,3)	(1,2,4)			15
(1,3,3)					3
(2,2,2)	(2,2,3)				4
				total =	35

This can be easily checked from Table 2.4–3.

These 11 are all the canonical cases that occur; we have only looked at about 5% of the 216 possible cases! We now multiply each canonical representation by its correct factor to get the number of equivalent representations out of the total 216 representations; as before if all three entries are the same then the factor is 1, if two are the same and one different then the factor is 3, and if all three are different then the factor is 6. We get the totals of each line on the right. The total number of possible outcomes is $6^3 = 216$, and each outcome has the same probability 1/216. The probability that the sum is not more than 7 is therefore $35/216 = 0.162\ldots$.

Backtracking has clearly eliminated, in this case, a lot of the sample space as a place to search for successes. The use of the canonical representations reduced the initial number of cases from 35 to 11. In many problems backtracking, and similar methods from AI, can greatly simplify the search of the total sample space for the sucesses.

Exercises 3.3

3.3–1 If on rolling two dice you win your point when either die has the number or when their sum is your number, then what are the probabilities for winning for each point (1 through 12)? (Cardan)

3.3–2 Extend the gold coin problem of Example 1.9–1 and Exercise 3.2–3 to n coins in $n+1$ drawers. Ans. 2/3.

3.3–3 Extend Exercise 3.3–2 to drawing k gold coins and then estimating the probabilty of the next being gold. Ans. $(k+1)/(k+2)$.

3.3–4 Find directly the probability of the sum of three dice being less than or equal to 8. Check by Table 2.4–3.

3.3–5 Find the probability of the sum of four dice being less than or equal to 7

. **3.3–6** What is the probability of getting either a 7 or an 11 in the roll of two dice? Ans. 2/9

3.3–7 What is the probability of getting a total of 7 on the roll of four dice?

3.3–8 Using back tracking find a position on a chess board for which there are 8 queens each of which is not attacked by any other queen.

3.3–9 Find all such board positions asked for in Exercise 3.3–8.

3.3–10 If the face cards count $J = 11$, $Q = 12$, and $K = 13$ what is the probability of two random draws (with replacement) totaling more than 15?

3.3–11 Same as 3.3–10 except without replacement.

3.3–12 In the toss of 6 coins what is the probability of more than 3 heads?

3.3–13 In a hand of 13 cards four are aces. What is the probable number of face cards?

3.3–14 In Exercise 3.3–13 what is the probability of no face cards? Ans. $C(36,9)/C48,9)$

3.4 Historical Approach

Early in the correspondence between Fermat and Pascal (which is often said to have begun the serious development of probability theory) *the problem of points* was discussed. The question was: in a fair game of chance two players each lack sufficient points in order to win. If they must separate without finishing the game, how should the stakes be divided between them?

Example 3.4–1   *The Problem of Points*

The *historical approach*, this time used *backwards*, provides a convenient approach. We draw figure 3.4–1 for the case $n = 3$ as the number of points

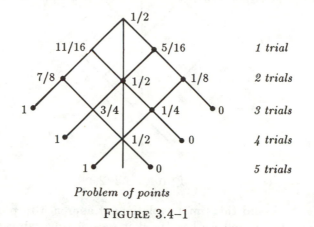

Problem of points

FIGURE 3.4–1

needed to win. The figure is the complete "tree" starting at the top with no trials made, and going to the left for each success of player A and to the right for a failure, each with the assumed probability of 1/2. At each level we have the situation after the indicated number of trials. The nodes that end with a win for A are marked with a payment of 1, and those for B are marked with 0, meaning that A gets nothing. We then "back up" the tree (as with a PERT chart) to get the values to assign earlier values (the average of the two descendant nodes). These derived values are marked on the nodes and give the value of the game to A at that position. At a symmetrical point (about the vertical middle line) the value for B is the complement (with respect to 1) of the marked value. The values for B are naturally the complements of those for A.

Example 3.4–2   *Three Point Game With Bias*

Suppose that in the game of points the probability is not 1/2 but rather A has a probability p of winning at each trial, and that the play is supposed to continue until one player has a total score of 3.

We again draw a "tree," Figure 3.4–2, and mark the wins and losses as before. But this time when we "back up" the values we must consider the probabilities of the events. We get the indicated values for the game at each stage, including the interesting value at the top for the advantage A has at the start of the game. The coefficients of the terms at each node are actually the number of paths for reaching a win from that position, and the powers of p and q in the term are the number of steps in the corresponding directions.

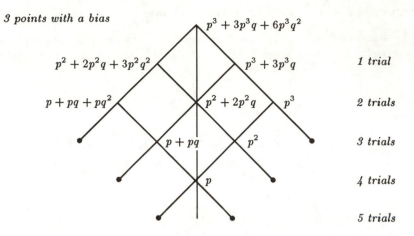

FIGURE 3.4-2

This (backward this time) evolutionary approach to probability problems is very useful, and we have used it several times already. The above example shows a slight twist to the concept, examining how the winning positions can arise, and hence their probabilities.

Example 3.4-3   *Estimating the Bias*

If in the game of points one is forced to stop before the end, and if the outcome of a trial *depends on skill* rather than luck, then the ratio of successes to the total number of games so far played will give an indication of the probability p of a given player winning at a single trial. This has, of course, all the risk of any statistical estimate, but if the game depends on skill and you have no other data available to estimate the relative skills of the two players (at that moment in time), then what else is reasonable? Thus the ratio of games so far won to the total tried gives the estimates of the p to be used.

We have used the historical approach a number of times before, for example in the birthday problem, so more examples are not needed now.

Exercises 3.4

3.4–1   Show that Figures 3.4–1 and 3.4–2 agree.

3.4–2   Discuss interchanging p and q in Figure 3.4–2.

3.4–3   Work out the details for Example 3.4–3.

3.4–4   If there are 5 white balls and 4 black balls in an urn, what is the probability that on drawing them all out they will be alternating in color? Ans. $p = 1/126$.

3.4–5   In Example 3.4–3 show how the proposal tends to make small differences in ability into greater differences.

3.4–6 There are 6 discs numbered 1, ..., 6. What is the probability of drawing them in a preassigned order? Ans. 1/6!

3.4–7 In drawing the six discs in the previous problem what is the probability that you will get all the even numbered ones before the odd numbered ones?

3.4–8 Previous problem with $2n$ discs. Ans. $(n!)^2/(2n)!$

3.4–9 In a bag of 2 red, 3 white, and 4 black balls what is the probability that you will get all the red, then the white, and finally the black balls in that order?

3.5 Recursive Approach

The recursive method of solving probability problems is very powerful, and often gives both: (1) valuable cross checks, and (2) a feeling for why the particular case comes out as it does. It is, in a sense, the historical approach with a great deal of regularity.

Example 3.5–1 *Permutations*

The simplest example of the recursive method is the derivation of the number of permutations. If $P(n)$ is the number of permutations of n things, then clearly you can pick any one of the n items and have left $n-1$ items,

$$P(n) = nP(n-1) \qquad P(1) = 1$$

We solve this recursion (difference equation) as follows:

$$P(n) = n(n-1)P(n-2) = n(n-1)(n-2)P(n-3) = \cdots = n!$$

Our attention is focused at just two places, the recurrence relation and the intital condition; we do not need to repeat the same argument again and again.

Example 3.5–2   *Runs of Heads*

What is the probability of a run of at least n heads? We have for a run of one head the probability p. This is a basis for the recursive approach. We assume that we have the probability $P(n)$ of a run of n heads and ask for the probability of a run of $n + 1$ heads. It is clearly given by

$$P(n + 1) = pP(n), \qquad P(1) = p \quad \text{or} \quad P(0) = 1$$

whose solution (see Appendix 4.A) is $p(n) = p^n$ (using the first case to determine the arbitrary coefficient of the solution).

If the run is to stop at exactly n heads then the next outcome must be a tail and we have

$$p^n q$$

as the probability of exactly n heads. The sequence $\{p^n q\}$ is the distribution of exactly n successes.

In Example 3.2–2 we examined the probability of getting two or no heads with two biased coins. In Exercise 3.2–4 we examined the probability of three or one heads in three tosses of a biased coin. Hence we next look at the general case of N tosses of a coin and getting exactly N heads or less than N by an even number. This problem again illustrates the recursive method.

Example 3.5–3   *N Biased Coins*

Given N biased coins, $p_i(H) = 1/2 + e_i$, what is the probability P_N of the number of heads being either N or less than N by an even number? We solve it, naturally, by a recursive method. We have the cases $N = 1, 2$, and 3 as a basis for the induction on the number of coins N. We have the recurrence equation describing a success at stage N based on the success or failure at stage $(N - 1)$ coins,

$$P_1 = 1/2 + e_1$$

$$P_N = P_{N-1}(1/2 + e_N) + (1 - P_{N-1})(1/2 - e_N)$$

$$= (1/2)P_{N-1} + P_{N-1}e_N + 1/2 - (1/2)P_{N-1} - e_N + P_{N-1}e_N$$

$$= (1/2) - e_N + 2P_{N-1}e_N$$

If we set

$$P_k = 1/2 + E_k$$

where E_k is the bias at stage k of the induction (note $E_1 = e_1$). We have from the above equation

$$1/2 + E_N = 1/2 - e_N + 2[1/2 + E_{N-1}]e_N$$

from which we get for the bias at stage N

$$E_N = 2e_N E_{N-1}$$

The solution of this recurrence relation is

$$E_N = 2^{N-1} e_N e_{N-1} \ldots e_1$$

hence

$$P_N = 1/2 + 2^{N-1} e_1 e_2 \ldots e_N$$

This solution checks for $P_0 = 1$, $P_1 = 1/2 + e_1$, and the previous results.

Note 1. If any one coin is unbiased, (some $e_k = 0$), then no matter how biased the other coins are (even if all the others are two headed or two tailed) *the game is fair.*

Note 2. If you change the rule to "an even (odd) number of heads" the answer changes at most by minus signs only.

Note 3. As long as some fraction of the coins have $|e_k| < e < 1/2$ then $P_N \to 1/2$ as $N \to \infty$. Notice that this does *not* involve the weak law of large numbers.

Example 3.5–4   *Random Distribution of Chips*

Suppose there are a men and b women, and that n chips are passed out to them at random. What is the probability that the total number of chips given to the men is an even number?

We regard this as a problem depending on n. We start with $n = 0$. The probability that the men have an even number of chips when no chips have been offered is (since 0 is an even number)

$$P(0) = 1$$

How can stage n arise? From stage $n - 1$, of course. We have (similar to the previous example) the recurence equation

$$P(n) = [b/(a + b)]P(n - 1) + [a/(a + b)][1 - P(n - 1)]$$

We write this difference equation in the cannonical form (see Appendix 4.A)

$$P(n) - \frac{b - a}{b + a} P(n - 1) = \frac{a}{b + a}$$

For the moment we write

$$(b - a)/(b + a) = A$$

The difference equation is now

$$P(n) - AP(n-1) = \frac{a}{b+a}$$

The *homogeneous* difference equation

$$P(n) - AP(n-1) = 0$$

has the solution (it can also be found by simple recursion)

$$P(n) = CA^n$$

for some constant C.

For the particular solution of the complete equation we try an unknown constant B as a guessed solution. We get

$$B - AB = \frac{a}{b+a}$$

or

$$B = \left(\frac{a}{b+a}\right)\frac{1}{1-A} = \left(\frac{a}{b+a}\right)\left[\frac{a+b}{a+b-(b-a)}\right]$$

$$= a/2a = 1/2$$

Thus the general solution of the difference equation is

$$P(n) = 1/2 + C\left\{\frac{b-a}{b+a}\right\}^n$$

We now fit the initial conditions $P(0) = 1$. We get immediately that $C = 1/2$, and the solution is

$$P(n) = \frac{1}{2}\left[1 + \left\{\frac{b-a}{b+a}\right\}^2\right]$$

$$= \frac{1}{2}\left[\left\{\frac{b+a}{b+a}\right\}^n + \left\{\frac{b-a}{b+a}\right\}^n\right]$$

which is more symmetric, and perhaps easier to understand.

To check the solution, we note that if $\mathbf{a} = 0$, meaning there are no men, then certainly (meaning $P(n) = 1$) the number of chips given to them is an even number as the formula shows. If $\mathbf{a} = \mathbf{b}$ then we have for all $n > 0$

$$P(n) = 1/2$$

as it should from symmetry. Finally, if there are no women, $\mathbf{b} = 0$, then

$$P(n) = \frac{1 + (-1)^n}{2}$$

which alternates 0 and 1 as it should. Thus we have considerable faith in the solution obtained.

Example 3.5–5   *The Gambler's Ruin*

Suppose gambler A holds **a** units of money and gambler B holds **b** units. If the probabilty of A winning on a single Bernoulli trial is p and they play until one of them has all the money, $T = a + b$, (T = total money) what is A's chances of winning?

We begin with the observation that we are concerned with a probability $P(n)$ if A holds exactly n units *for all* n ($0 \leq n \leq T$) since all such states may arise in the course of the game. We set up the standard equation, which gives the probability of A being in the state of holding n units; namely by either winning one trial and going from holding $n - 1$ to holding n, or else by losing and going from holding $n + 1$ to holding n. The difference equation is clearly

$$P(n) = pP(n - 1) + qP(n + 1) \qquad (n = 1, 2, \ldots, T - 1)$$

We next need the end conditions. When $n = 0$ (the gambler has no money, $P(0) = 0$) we know that A is certain to lose—A has no money to play. For $n = T$ we know A is certain to win (B has no money to play) so $P(T) = 1$. Thus the end conditions are

$$P(0) = 0, \qquad \text{and} \qquad P(T) = 1$$

To solve the difference equation (in its standard form)

$$qP(n + 1) - P(n) + pP(n - 1) = 0$$

we naturally try $P(n) = r^n$. This gives the characteristic equation (provided $pq \neq 0$)

$$qr^2 - r + p = 0$$

whose roots are

$$r = \frac{1 \pm \sqrt{1 - 4pq}}{2q}$$

We first look at the important special case of $p = 1/2 = q$. For this we have multiple characteristic roots

$$r = 1, 1$$

and the solution of the difference equation is

$$P(n) = C_1 + C_2 n$$

From $P(0) = 0$ we get $C_1 = 0$, and from $P(T) = 1$ we get $C_2 = 1/T$. Hence the solution is

$$P(n) = n/T.$$

Correspondingly A's probabilty of ruin is $1 - n/T$. At the start of the game ($n = a$, the original amount of money he had) this probability of winning is

$$P(a) = \frac{a}{a+b}$$

Of course B's probabilties are the complements, (or interchange a and b).

We now return to the general case where p is not 0, 1/2, or 1. The term in the radical

$$1 - 4pq = 1 - 4p(1-p) = 1 - 4p + 4p^2 = (1-2p)^2$$

so the characteristic roots are

$$\{1 \pm (1-2p)\}/2q = 1, p/q$$

Hence the general solution is

$$P(n) = C_1 + C_2 \left(\frac{p}{q}\right)^n$$

We need to fit the boundary conditions. For $P(0) = 0$ we get

$$0 = C_1 + C_2 \Rightarrow C_1 = -C_2$$

and for $P(T) = 1$ we have

$$1 = C_2 \left\{ \left(\frac{p}{q}\right)^T - 1 \right\}$$

$$C_2 = 1 \Big/ \left\{ \left(\frac{p}{q}\right)^T - 1 \right\}$$

and

$$P(n) = \left\{ \left(\frac{p}{q}\right)^n - 1 \right\} \Big/ \left\{ \left(\frac{p}{q}\right)^T - 1 \right\}$$

At the start $n = a$ and $T = a + b$ and

$$P(a) = \left\{ \left(\frac{p}{q}\right)^a - 1 \right\} \Big/ \left\{ \left(\frac{p}{q}\right)^{a+b} - 1 \right\}.$$

The probability of ruin is the complement

$$Q(n) = 1 - P(n)$$

Alternately, A's probability of losing is exactly B's probability of winning.

For more material on the gambler's ruin, such as the probable length of play, see [F, p.345ff].

Exercises 3.5

3.5–1 Suppose the game in Example 3.5–4 is changed to an even number for a win. Find the sign to be properly attached at each stage N.

3.5–2 Check algebraically that the last statement in this section is correct.

3.5–3 Discuss the cases in Example 3.5–5 when $pq = 0$.

3.5–4 If in Example 3.5–4 there are twice as many women as men what is the answer? Is it reasonable? If there are k times as many?

3.5–5 Discuss the convergence of the solution in Example 3.5–4 as the number of chips grows larger and larger.

3.5–6 Discuss the "drunken sailor" random walk along a line starting at $x = 0$. What is the probable distance from $x = 0$ after N steps?

3.5–7 The gamma function is defined by the integral $\int_0^\infty x^{n-1} e^{-x} \, dx$ For integral n find the recurrence formula and the value of the integral. Ans. $(n-1)!$

3.5–8 Evaluate $L_n = \int_0^1 (\ln x)^n \, dx$ by a recursion method.

3.5–9 The Catalan numbers are defined by $C(2n, n)/(n+1)$. Find the recursion formula and evaluate the first five numbers.

3.6 The Method of Random Variables

We illustrate this important method of random variables by several examples. We have already in this Chapter, Section 3.1 and Example 3.2–1, used this method.

Example 3.6–1   *Pairs of socks*

Suppose there are n pairs of socks in a dryer and you pick out k socks. What is the average number of pairs of socks that you will have?

Let the random variable $X_{i,j}$ be defined by

$$X_{i,j} = \begin{cases} 1 & \text{if the } i^{\text{th}} \text{ and } j^{\text{th}} \text{ socks form a pair} \\ 0 & \text{otherwise} \end{cases}$$

Let the ith sock be drawn by chance; since there are now $2n - 1$ socks left the probability that the jth sock will form a pair with it is $1/(2n - 1)$, that is

$$P\{X_{i,j}\} = E\{X_{i,j}\} = \frac{1}{2n - 1}$$

Now let the random variable X be the sum

$$X = \sum_{1 \le i,j \le k} X_{i,j} \qquad (i \ne j)$$

hence we have

$$E\{X\} = \sum_{i,j} E\{X_{i,j}\} = \frac{C(k,2)}{2n-1}$$

$$= \frac{k(k-1)}{2(2n-1)}$$

We make a small table of the values of $E\{X\}$, both to check the formula and get a better understanding of how the problem depends on the parameters k and n.

TABLE 3.6–1

The Random Socks Problem

$k \backslash n$	$n = 1$	$n = 2$	$n = 3$
$k = 2$	$(2 \times 1)/(2 \times 1) = 1$	$(2 \times 1)/(2 \times 3) = 1/3$	$(2 \times 1)/(2 \times 5) = 1/5$
$k = 3$	$-$	$(3 \times 2)/(2 \times 3) = 1$	$(3 \times 2)/(2 \times 5) = 3/5$
$k = 4$	$-$	$(4 \times 3)/(2 \times 3) = 2$	$(4 \times 3)/(2 \times 5) = 6/5$
$k = 5$	$-$	$-$	$(5 \times 4)/(2 \times 5) = 2$
$k = 6$	$-$	$-$	$(6 \times 5)/(2 \times 5) = 3$

Many of the values in the table agree with plain thinking. For larger values, if $k = 2n$ (all the socks are drawn) we get

$$E(X) = \frac{2n(2n-1)}{2(2n-1)} = n$$

and we have n pairs of socks. If $k = 2n - 1$ (we leave exactly one sock) then we have

$$E(X) = \frac{(2n-1)(2n-2)}{2(2n-1)} = n - 1$$

pairs of socks. Hence we decide that the answer is probably right.

Example 3.6–2   *Problem de Recontre*

This famous matching problem may be stated in many forms. At a dance n men draw names of the wives at random; what is the probability that no man dances with his wife? Or n letters and envelopes are addressed, and then the letters are put in envelopes at random; what is the probabilty that no letter is in its envelope? Again, at an office party where each person puts a gift in a bag and later draws at random the gift, what is the chance that no one will get the gift they put in? These are the same question as the probability of no match when turning up cards from two well shuffled decks of cards and asking for the probability of no match. Indeed, it is the same as calling the cards in order and looking for no match from a well shuffled deck. There are many other equivalent versions of the problem.

Let the random variable X_i be defined by

$$X_i = \begin{cases} 1 & \text{if the } i^{\text{th}} \text{ letter is in the } i^{\text{th}} \text{ envelope} \\ 0 & \text{otherwise} \end{cases}$$

The random variable X we are interested in is defined by

$$X = \prod_{i=1}^{n}(1 - X_i)$$

which is 1 if no letter is in its envelope and 0 otherwise. We have, on expanding the product

$$X = 1 - \sum X_i + \sum X_i X_j - \sum X_i X_j X_k + \cdots$$

$$E\{X\} = 1 - n\{1/n\} + C(n,2)\{1/n(n-1)\} - C(n,3)\{1/n(n-1)(n-2)\}$$

$$+ \cdots$$

$$= 1 - 1 + 1/2! - 1/3! - \cdots + (-1)^n/n!$$

$$\sim 1/e = 0.367879\ldots$$

The probability of at least one letter in its proper place is

$$1 - E\{X\} \sim 1 - 1/e = 0.632120558\ldots.$$

The approach to the limiting value $1 - 1/e$ is very rapid as the following short table for at least one in the correct envelope shows;

TABLE 3.6-2

n	exact value
1	1.0
2	0.5
3	0.666667
4	0.625
5	0.633333
6	0.631944
7	0.631429
8	0.632118
9	0.632121
10	0.632121

Thinking about the table you soon realize that the larger n is the easier it is for any one letter to miss its envelope, but the more letters give more chances for at least one hit; the table shows that in this case the two effects almost exactly cancel each other, once n is reasonably large.

Example 3.6–3   *Selection of N Items From a set of N*

Evidently if we did not replace an item after drawing it then after N draws we would have the complete set. But with replacement after each draw we can anticipate getting some items several times and hence missing some others. How many can we expect to miss?

Let the random variable X_i be

$$X_i = \begin{cases} 1 & \text{if the } i^{\text{th}} \text{ item is not taken} \\ 0 & \text{if the } i^{\text{th}} \text{ item is taken} \end{cases}$$

Therefore defining $X = X_1 + X_2 + \cdots + X_N$

$$E\{X\} = \sum E\{X_i\}$$

But the X_i are interchangable, hence

$$E\{X\} = N E\{X_i\}$$

Now what is the probability of missing the ith item in the N trials? If each trial must miss then we have

$$[(N-1)/N]^N = [1 - 1/N]^N \sim 1/e = 0.3678\ldots$$

(see Table 2.4–2). Hence the expected number of misses is

$$N/e$$

and we miss getting about 1/3 of the items when replacing each item immediately after each trial.

Exercises 3.6

3.6–1   Explain the oscillation in Table 3.6–2 in terms of the problem posed.

3.6–2   Give another example of the application of Example 3.6–2.

3.6–3   There are n sets of three matching items thrown in a bag. If you draw them out one at a time what is the expected number of matching sets?

3.6–4   In sampling n times with replacement what is the probability of getting all the items from a collection of n when is $n = 1, 2, 3, \ldots, n$?

3.7 Critique of the Notion of a Fair Game

Galileo was once asked, "In estimating the value of a horse, if one man estimated the value at 10 units of money and the other at 1000 units, which of two made the greater error in estimating the value if the true value is 100 units of money?" After some consideration Galileo said that both were equally accurate, that it is the *ratio* of the estimate to the true value, rather than the difference, that is important.

To put the matter on a more personal level, suppose you have 1 million dollars and stand to either win or lose that amount on an even bet ($p = 1/2$). Most people do not feel that the gain in going from 1 million to 2 million dollars is as great as is the loss in going from 1 million dollars to nothing.

Thus we see that the definition of a fair game involving money, where the expectation of the loss is equal to the expectation of gain, does *not* represent reality for most people. Bernoulli faced this matter and suggested that the log of the amount you have after the gain or loss relative to what you had originally, is more realistic—but the $\log 0 = -\infty$ and this may seem to be a bit too strong (so you might use the log of a constant plus the amount). Yet the log is not a bad measure to use in place of the amount itself. In any case we see that, except for very small amounts (small relative to the total amount possessed), the definion of a fair game is often *not* appropriate to life as we lead it. It has nice mathematical properties, that is true, but to apply it without thought of its implications is foolish; the concept needs to be watched very carefully when it is used in various arguments.

However, there is another side to the argument. People who engage in State Lotteries, when asked why they accept such unfavorable odds, give arguments (if they reply at all) such as, "Buying a ticket a week and losing each time will hardly change my life style, but if I win then it will make a great deal of difference to me." Thus they feel that the slight loss in the expected value of their life style that they will suffer if they always lose is more than compensated for by the possibility (not the probability) that they will win a large amount on the lottery. They also often believe in a "personal luck" and not in the kinds of dispassionate, impersonal, objective probabilities that are expounded in this book.

One may also say about the usual concept of insurance, for the person taking out the policy there is a slight loss in the expected quality in their life (due to the premium paid) but a greatly reduced possible loss. Gambling may be viewed as the opposite of insurance, for a moderate loss in the expected value, the possible gain in their life is greatly raised. These opinions are to be considered more on an emotional level than on a rational level since we cannot know the nonmeasureable values attributed to them by the individual making the choice.

As we will show in Section 3.8, using the Bernoulli evaluation, the log of the amount of money posessed, leads to the reasonable results that gambling in fair games is foolish, and that ideally (no overhead) insurance is a good idea. Bernoulli used the log of the amount as the evaluation function, but it will be easy to see that any reasonable convex function would give similar results.

While we have just argued that the Bernoulli value of money, or some similar function, is the proper one to use when dealing with money, it does not follow that for other quantities, such as time, it is appropriate. In each case we must examine the appropriateness of using the expected value as the proper measure before acting on a computation involving the expected value.

3.8   Bernoulli Evaluation

In this section we give several examples of the Bernoulli evaluation of a fair game.

Mathematical Aside: *Log Expansions*. We often need the following simple results. From the expansion (that can be found by simple division if necessary)

$$1/(1-x) = 1 + x + x^2 + x^3 + \cdots \qquad (|x| < 1)$$

we can integrate (from 0 to x) to get (recall that $ln 1 = 0$)

$$ln(1-x) = -(x + x^2/2 + x^3/3 + \cdots) \qquad (3.8\text{--}1)$$

Replace x by $-x$ to get

$$ln(1+x) = x - x^2/2 + x^3/3 - \cdots \qquad (3.8\text{--}2)$$

Subtract (3.8–1) from (3.8–2) to get

$$ln\{(1+x)/(1-x)\} = 2(x + x^3/3 + x^5/5 + \cdots) \qquad (3.8\text{--}3)$$

You can, if you wish, use finite expansions with a remainder in all the above, and finally go to the limit; the result will be the same.

Example 3.8–1   *Coin Toss*

Let $p = 1/2$ for a fair game with a gain or loss of one unit of money on the toss. Then the value of the game according to Bernoulli is, if you have N units of money at the start, (V_B = value Bernoulli)

$$V_B = \tfrac{1}{2}ln(N+1) + \tfrac{1}{2}ln(N-1)$$

The change in your value (you had $ln\ N$ value before the game) by the toss is

$$\Delta V_B = \tfrac{1}{2}[ln(N+1) + ln(N-1) - 2ln(N)]$$

$$= \tfrac{1}{2}[ln(1+1/N) + ln(1-1/N)]$$

$$= \tfrac{1}{2}ln(1-1/N^2) < 0$$

Hence by (3.8–1) there is approximately a change of $-1/2N^2$ in your value, and it is foolish (in an economic sense) to play the game.

From the previous example we can fairly easily see that if the payoff is symmetric in a fair game then each matching pair in the payoff is unfavorable, hence repeated trials are also unfavorable, and playing the game repeatedly is foolish. We show this in more detail in the next Example.

Example 3.8–2   *Symmetric Payoff*

For the toss of three coins (the order of the coins does not matter) let the payoff be the number of heads minus the number of tails, and suppose you start with N units of money.

outcome	payoff	probability
HHH	$3A$	$1/8$
HHT	A	$3/8$
HTT	$-A$	$3/8$
TTT	$-3A$	$1/8$

Is the game fair both in the classical sense and in the Bernoulli sense?

The game is clearly fair in the classical sense. However, the change in the Bernoulli value is

$$change = \tfrac{1}{8}[ln(N+3A) + ln(N-3A)]$$

$$+ \tfrac{3}{8}[ln(N+A) + ln(N-A)] - lnN$$

$$= \tfrac{1}{8}[ln\{1 - (3A/N)^2\} + 3ln\{1 - (A/N)^2\}]$$

and by 3.8–1 this is negative.

We are now ready to look at the general case of a fair game.

Example 3.8–3   *Fair Games*

If you have N units of money and

$$p = \text{probability of winning } A \text{ units}$$

$$q = \text{probability of losing } B \text{ units,} \qquad (B < N)$$

then for a classic fair game you must have

$$pA = qB = C$$

and the net gain (Bernoulli) is

$$\text{Gain } (N) = G(N) = p \, ln(N + A) + q \, ln(N - B) - ln \, N$$

$$= p \, ln(1 + A/N) + q \, ln(1 - B/N)$$

To study the function $G(N)$ as a function of N we note that $G(\infty) = 0$ regardless of the other parameter values p, q, A, and B. Next we examine the derivative with respect to N

$$G'(N) = p \left[\frac{\dfrac{-A}{N^2}}{1 + \dfrac{A}{N}} \right] + q \left[\frac{\dfrac{B^2}{N}}{1 - \dfrac{B}{N}} \right]$$

$$= \frac{1}{N} \left[\frac{-C}{N + A} + \frac{C}{N - B} \right]$$

$$= \frac{C}{N} \left[\frac{1}{N - B} - \frac{1}{N + A} \right]$$

$$= \frac{C}{N} \left[\frac{A + B}{(N - B)(N + A)} \right] > 0$$

The derivative is always > 0, hence the curve must be rising, and since it is 0 at infinity we must have, for all finite N,

$$G \text{ (finite) } < 0 \qquad (\text{for } N > B)$$

Example 3.8–4   *Insurance (Ideal—no Extra Costs)*

In the insurance case, you are already in the game of life and you stand to lose an amount $A < N$ (Where $N = $ your total assets) with some probability p. In this game of life your current Bernoulli value is

$$p \ln(N - A) + q \ln N$$

If you took out insurance so that you would get back the amount A, then a (classical) fair fee would be (ideally)

$$pA$$

and your Bernoulli value would then be

$$\ln(N - pA)$$

Computing the gain in taking the insurance

$$(\text{insurance}) \; - \; (\text{no insurance})$$

you get the gain

$$G(N) = \ln(N - pA) - [p \ln(N - A) + q \ln N]$$

$$= \ln(N - pA) - p \ln(N - A) - (1 - p)\ln N$$

$$= \ln(1 - pA/N) - p \ln(1 - A/N)$$

Again we observe that

$$G(\infty) = 0$$

but this time

$$G'(N) = \frac{\dfrac{pA}{N^2}}{1 - p\dfrac{A}{N}} - \frac{\dfrac{pA}{N^2}}{1 - \dfrac{A}{N}}$$

$$= \frac{pA}{N}\left[\frac{1}{N - pA} - \frac{1}{N - A}\right]$$

$$= \frac{pA(-qA)}{N(N - A)(N - pA)} < 0$$

Hence the curve always falls as a function of N, and therefore the function values must have been positive. Thus taking insurance is always good (assuming no overhead and that the insurance is priced fairly!).

Since there are actually fixed overheads in insurance policies the results must be slightly modified. It is easy to see that if your N is large it does not pay you to insure against a small loss, but when the loss is near to N is generally pays to take insurance. As a result the richer people tend to have fewer insurance policies than do the poorer people.

An examination of the last two proofs shows that any reasonable convex function that is differentiable in place of the log would not change the results. Hence the Bernoulli (log) value of money is not essential to the conclusions. The conclusion is that the classical "fair game" involving money is simply not "fair" will still hold for convex value functions. Thus the wide spread use of the expected value in probability can, at times, be grossly misleading.

Exercises 3.8

3.8–1 For the roll of a die if the payoff is 6 units for "hitting your point" show that the game is fair but the Bernoulli evaluation is not.

3.8–2 For the toss of a coin twice with (in order) HH payoff $2A$, HT payoff A, TH payoff $-A$ and TT payoff $-2A$, show that the game is fair but the Bernoulli evaluation is not.

3.8–3 What would the payoff have to be in a coin toss for the Bernoulli value to be zero? Note that it involves your present value.

3.8–4 Can a game be both fair and have a Bernoulli evaluation of zero?

3.8–5 Calculate the Bernoulli value of $2n$ fair games.

3.8–6 Replace the $\log N$ value with $\log(N + 1)$ and carry out the details, thus avoiding the infinite value when ending up with nothing.

3.9 Robustness

As you have seen, the first modelling of a situation often asumes that the probability distribution is uniform. It may in fact be only close to uniform distribution, so it is natural to ask how much the answer changes and in what direction it changes if there are slight differences from uniform in the distribution assumed. This is usually called "robustness" of the model, or "sensitivity" if you prefer. If we are to use the results of a probability computation in many practical situations we must not neglect this important aspect of practical mathematics. We will take the birthday problem of Example 1.9–6 as an example.

Example 3.9-1   *The Robust Birthday Problem*

Suppose that the distribution of birthdays in the year is not uniform, but assume that the probabilities deviate from 1/365 by small amounts e_i, that is (still using 365 days in the year and neglecting leap years) the probability of being born on the ith day of the year is

$$p_i = 1/365 + e_i \qquad (i = 1, 2, \dots 365) \qquad (3.9\text{-}1)$$

For generality we replace 365 by an arbitrary n.

We have, of course,

$$\sum_i e_i = 0 \qquad (3.9\text{-}2)$$

Therefore

$$\left(\sum_i e_i\right)^2 = 0$$

by expanding the square of the series and transposing the squared terms we have the important relationship

$$\sum_{i \neq j} e_i e_j = -\sum_i e_i^2 \qquad (3.9\text{-}3)$$

Now the probability we seek P_m for no duplicate birthdays in m choices must be a symmetric function in the e_i and consist of all terms of the form

$$p_{i(1)} p_{i(2)} \cdots p_{i(m)}$$

such that $i, j \neq i, k$, for $j \neq k$. We make a Taylor expansion in the variables e_i of this function $P_m(p_i)$ about the values $e_i = 0$. We have

$$P_m = P_m(1/n) + \sum dP_m/dp_i \bigg|_{1/n} e_i$$

$$+ \sum \sum d^2 P_m/dp_i dp_j \bigg|_{1/n} e_i e_j + \cdots$$

We now note that due to symmetry all the

$$dP_m/dp_i \bigg|_{1/n} = \text{same number for all } i$$

hence

$$\sum dP_m/dp_i \bigg|_{1/n} e_i = dP_m/dp_i \bigg|_{1/n} \sum e_i = 0$$

by (3.9-2).

For the next term, the second derivatives evaluated at $1/n$, we have for $i \neq j$,

$$d^2 P_m / dp_i dp_j \Big|_{1/n} = \begin{cases} 1/n^{m-2} & \text{if both } p_i \text{ and } p_j \text{ are there} \\ 0 & \text{otherwise} \end{cases}$$

and that

$$d^2 P_m / dp_i^2 = 0 \text{ always}$$

Hence

$$P_m(p_i) = P_m(1/n) + (1/n^{m-2}) \sum e_i e_j + \cdots$$

Using (3.9–3) we get, finally,

$$P_m(p_i) = P_m(1/n) - (1/n^{m-2}) \sum e_i^2 + \cdots$$

$$= (\text{original answer}) - (1/n^{m-3}) \text{Var}\{e_i\} + \cdots$$

and we have found the major correction term in a very convenient form.

The equations (3.9–2) and (3.9–3) are the key ones in the reduction to the simple form; for such a symmetric problem the first order term will always cancel out, and the cross derivatives can sometimes be reduced to the second order ones and then the variance will emerge as the measure of the main correction term. We have, therefore, a convenient way of attacking the robustness of many symmetric problems when small deviations from the uniform distribution occur. For other kinds of problems we will need other tools, but equations (3.9–2) and (3.9–3) remain central.

The birthday problem is one of sampling with replacement; we next look at the elevator problem (Section 3.1) which samples without replacement and has a double index probability distribution.

Example 3.9–2   *The Robust Elevator Problem*

Looking back at the first solution (or the second one) we can take the six successes from the sample space

$$
\begin{array}{ccc}
a1 & b2 & c3 \\
a1 & b3 & c2 \\
a2 & b1 & c3 \\
a2 & b3 & c1 \\
a3 & b1 & c2 \\
a3 & b2 & c1
\end{array}
$$

and compute their total probability as the probability of a success. Let the probability of the ith person getting off on the jth floor be

$$p(i, j) \qquad \text{with} \qquad \sum_{j=1,2,3} p(i,j) = 1 \qquad (\text{for } i = 1, 2, 3)$$

This sum merely says that the ith person gets off on some floor.

For use as perturbations we set

$$p(i,j) = 1/3 + e(i,j) \qquad \text{(all } i \text{ and } j)$$

with

$$\sum_{j=1,2,3} e(i,j) = 0 \qquad (i = 1,2,3) \qquad\qquad (3.9\text{-}4)$$

The total probability is the sum of six terms of the form

$$p(1,i)p(2,j)p(3,k) = [1/3 + e(1,i)][1/3 + e(2,j)][1/3 + e(3,k)]$$

To get the Taylor series expansion of the sum we need the derivatives with respect to the $e(i,j)$ evaluated at $e(i,j) = 0$. We have

$$P(0) = 6/27 = 2/9$$

$$\frac{dP(0)}{de(i,j)} = A = 1/9$$

$$\frac{d^2 P(0)}{de(i,j)de(k,m)} = B = 1/3 \qquad (i \neq k, j \neq m)$$

$$\frac{d^2 P(0)}{de(i,j)^2} = 0$$

The Taylor expansion is, therefore, since again the first order terms must cancel,

$$P(e) = \frac{2}{9} + \frac{1}{2}\left(\frac{1}{3}\right) \sum_{i \neq k, j \neq m} e(i,j)e(k,m)$$

If we sum on the m for fixed i, j, k we have from (3.9-4)

$$\sum_{m \neq j} e(i,j)e(k,m) = -e(i,j)e(k,j)$$

as a partial simplification for the expression. We can also write it in the form

$$P(e) = \frac{2}{9} + \frac{1}{6}\{\sum e_{i,j}e_{k,m} - \sum e_{i,j}^2\}$$

But we cannot reduce it to a sum of squares as the following argument shows. Consider the case $p(1,1) = p(2,2) = p(3,3) = 1$. Then $P = 1$, success is certain. Next consider the case, $p(1,1) = 1 = p(2,1)$. Then $P = 0$, and failure is certain. Thus the surface in the perturbation variables $e(i,j)$ has a saddle point (as can be seen by considering slight changes towards the two solutions indicated). Still, we have a formula for the change due to small changes in the probabilites (we have neglected any correlations that might also occur in practice). We see that again the change is quadratic in the perturbations, as one would expect from local extreme in a symmetric problem. But see [W].

We gave a rather general approach to robust problems and illustrated it by the birthday problem in Example 3.9–1. We now give an approach based on a variant of the birthday problem.

Example 3.9–3   A Variant on the Birthday Problem

Let N be the number of people entering the room, one at a time, until you have a 50% chance of a duplicate birthday. This differs from the earlier version of the birthday problem not only because it poses a different question but it also differs in the amount of effort to do a simulation. The question asked concerns the median value which is often quite different from the expected value.

To handle this problem this time we define the random variable X_j (for $0 < j < 366$)

$$x_J = \begin{cases} 1 & \text{when } N > j \\ 0 & \text{otherwise} \end{cases}$$

Consequently the number of people N until we get as 50–50 chance of a duplicate birthday is given by

$$N = X_0 + X_1 + \cdots + X_n$$

Next we take the expected value

$$E\{N\} = 1 + p_1 + p_2 + \cdots + p_n$$

where, as before,

$$p_j = \left\{ 1 - \frac{0}{365} \right\} \left\{ 1 - \frac{1}{365} \right\} \cdots \left\{ 1 - \frac{j-1}{365} \right\}$$

Using a programmable hand computer we get $E\{N\} = 24.617\ldots$

We next examine the variance of N in this formulation of the question. Note that since $X_i^2 = X_i$,

$$N^2 = X_0^2 + X_1^2 + \cdots + X_n^2 + 2 \sum_{i<j} X_i X_j$$

$$= X_1 + X_2 + \cdots + X_n + 2 \sum_{i<j} X_i X_j$$

Now the random variables were cleverly chosen so that for $i < j$

$$X_i X_j = X_j$$

Therefore

$$N^2 = X_0 + X_1 + \cdots + X_n + 2\sum_{i=0}^{n-1}\sum_{j=i+1}^{n} X_j$$

and it follows that

$$E\{N^2\} = E\{N\} + 2\sum_{i=0}^{n-1} r_i \qquad \text{where} \qquad r_i = p_{i+1} + \cdots + p_n$$

This formula also provides a way of computing when the probabilities of birthdays are not uniform over the year.

As an example of unequal probabilities of a birth on the various days of the year let $p_i = 1/365 + A\sin\{(k-1)2\pi/365\}$ with $k = 1, 2, \ldots 365$. This is a reasonable model if you think there is a periodic effect in the course of the year. With $A = 0.002$ (note that $1/365 = 0.00274$) this is a huge amplitude oscillation which produces an approximate variation of 73% in some of the daily probabilities. The result of computation is that the expected number is now

$$E\{N\} = 24.6$$

(which differs from the earlier uniform case only slightly. We have also

$$E\{N^2\} = 754.6166\ldots \qquad \text{Var}\ \{N\} = 148.65$$

hence

$$r = 12.19\ldots$$

The effect of the large change from the uniform case is not as much as you might expect because the sum of the perturbations in the probabilities must total 0 and this produces, in almost all probability problems, a great deal of cancellation.

Example 3.9–4   A Simulation of the Variant Birthday Problem

The result of Example 3.9–3 may be checked by some simulations of the situation. The average number N over 100 trials gave the numbers

20.97   21.14   23.16   22.48   22.20   23.59
20.07   21.79   23.21   20.63   22.32   22.33

The average of these simulations is 22.0 vs. the calculated 24.6 but the variance is large.

3.10 Inclusion–Exclusion Principle

We first introduce the concept of an *indicator variable* on a set X. The variable is 1 when the point is in the set and is 0 otherwise. Thus for indicator variables X_i the quantity

$$X_i X_j$$

is 1 only when the point is in both sets X_i and X_j. Similarly for higher order products of indicator variables. We have used this notation before.

Suppose we have a number of sets A_i, some of which may overlap others. If we add all the A_i we will at least double count the common points, labeled $A_{i,j}$, see Figure 1.7–2. Hence we need to subtract these out. But then we will have twice removed all the points that occur in three sets at once and we need to add them back again, that is add $A_{i,j,k}$. Hence to count all the points in some set we have

$$A = \sum A_i - \sum A_{i,j} + \sum A_{i,j,k} - \cdots$$

Figure 1.7–2 illustrates this situation for three sets using circles as sets. We now convert this to the indicator variables where where there is at least one point in the set X

$$X = \sum X_i - \sum X_i X_j + \sum X_i X_j X_k - \cdots$$

Since combinations are often difficult for the beginner, we will proceed slowly and derive this formula in a different manner. We begin with the classical formula from algebra, the factoring of a polynomial in terms of the roots x_i. We have the standard formula

$$\text{Poly } (x) = P_n(x) = (x - x_1)(x - x_2) \ldots (x - x_n)$$

If we multiply this out and note that the terms in each coefficient of x^k are homogeneous (same degree) in the x_i we find that the term with no x_i is

$$x^n$$

The next term is x^{n-1} and the coefficient is, by symmetry,

$$-\sum_i x_i$$

The next term's coefficient is

$$+\sum_{i,j} x_i x_j$$

Indeed we see that $P_n(x)$ will be

$$P_n(x) = x^n - x^{n-1} \sum x_i + x^{n-2} \sum x_i x_j - x^{n-3} \sum x_i x_j x_k + \cdots$$

with alternating signs and the last term will be the product of all the x_i with a coefficient $(-1)^n$. The coefficients are called the *elementary symmetric functions* of the roots x_i.

If we set $x = 1$ and think of the x_i as indicator variables X_i we have

$$(1 - X_1)(1 - X_2)\ldots(1 - X_n)$$

If any $X_i = 1$ then the product is 0. Alternately for

$$1 - (1 - X_1)(1 - X_2)\ldots(1 - X_n)$$

this is 1 if any $X_i = 1$. Expanding this we have

$$1 - 1 + \sum X_i - \sum X_i X_j + \cdots = \sum X_i - \sum X_i X_j + \sum X_i X_j X_k - \cdots$$

this is the same formula as before in a slightly different notation, and we have already used it several times!

Example 3.10–1   *Misprints*

A common situation is that two readers independently examine a table of numbers. The first observes n_1 errors, the second n_2 errors, and in comparing the two lists of errors there are n_3 errors in common. How many errors are probably left?

We must, of course make some assumptions. The simplest are: (1) that each error has the same probability of being found as any other, (2) that the errors are independent, and (3) that the two readers have different probabilities of finding the errors. Let $p(1)$ be the probability of the first reader finding an error and $p(2)$ for the second. If there are N errors in total then reasonable estimates (not certain knowledge at all) are

$$p(1) = n_1/N \qquad \text{and} \qquad p(2) = n_2/N$$

We have for the common errors (they were assumed to be independent)

$$p(1)p(2) = n_3/N$$

This leads to, eliminating the p's,

$$n_1 n_2 = N n_3 \Rightarrow N = n_1 n_2/n_3$$

This seems to be a reasonable result as we inspect it. If there were no errors found in common, (and nothing else was known about the reliability of the readers) the assumption of an infinite number of errors (though the text must have been finite) is not surprising. Other tests also show that it is a reasonable result for a first guess.

We have not answered the original question—how many are left. We need the inclusion–exclusion principle to get it. We have for the number of errors left the reasonable guess

$$N - n_1 - n_2 + n_3$$

The model is crude, of course, but it is suggestive, and if this number is too large for publication then further readers had better be found.

Example 3.10–2   *Animal Populations*

It is a common problem that we want to sample an animal (or other) population and guess at the total population N from the samples. Suppose we sample, count, and tag all the animals caught in this first sample, and return them to the original population. At a later date we sample again and count both the size and the number of tagged ones. Again we are estimating the total population (above it was errors in a table) and have labeled it as N. Let the n_i be as before. Then the total population N is estimated by

$$N = n_1 n_2 / n_3$$

One worries about the assumption of the independence of being captured twice. Are the trials really independent? Is there a tendency for some animals to be caught and other to escape? Does the presence of a tag change the probability of being caught again, or of survival to be later caught? Hence the need for robustness.

Example 3.10–3   *Robustness of Multiple Sampling*

The formula we are using is

$$N = n_1 n_2 / n_3$$

If each of these numbers has a small error e_i ($i = 1, 2, 3$) what can we expect for the change in N?

It is simple numerical analysis where we omit all squares and higher powers of the e_i. We have for the first order terms in the e_i

$$\text{Change} = (n_1 + e_1)(n_2 + e_2)/(n_3 + e_3) - n_1 n_2 / n_3$$

$$\sim [(n_1 n_2 + n_1 e_2 + n_2 e_1)/n_3]\{1 - e_3/n_3\} - n_1 n_2/n_3$$

$$\sim n_1 e_2/n_3 + n_2 e_1/n_3 - n_1 n_2 e_3/n_3^2$$

This could also be found by a Taylor expansion to first order terms.

Example 3.10–4   *Divisibility of Numbers*

A number is chosen at random from the range of $30M + 1$ to $30N(N > M)$; what is the probability that it will have no factors of 2, 3, or 5?

Clearly we first need to find how many numbers in the range are divisible by 2, 3 and 5. We have

divisible by 2 $30(N - M)/2 = 15(N - M)$

divisible by 3 $30(N - M)/3 = 10(N - M)$

divisible by 5 $30(N - M)/5 = 6(N - M)$

etc.

We must use the inclusion-exclusion principle, hence we have in all that are divisible by 2, 3 or 5

$$30(N - M)\left\{ \frac{1}{2} + \frac{1}{3} + \frac{1}{5} - \left\{ \frac{1}{6} + \frac{1}{10} + \frac{1}{15} \right\} + \frac{1}{30} \right\}$$

$$= (N - M)[15 + 10 + 6 - 5 - 3 - 2 + 1]$$

$$= (N - M)(22)$$

There are $30(N - M)$ numbers in the range, thus the probability of being divisible by 2, 3 or 5 is $22/30 = 11/15$, so that the probability of not being divisible is $4/15$.

Exercises 3.10

3.10–1 In Example 3.10-2 if the sampled fish die and cannot be returned alive (sampling without replacement) then find a formula for estimating the population left after the two samplings.

3.10–2 To estimate the number N of errors left in some software it is proposed to add M additional ones (of the same general type) and then search for the errors. If n of the original ones and m of the inserted ones are found, find a formula for estimating the number of errors left. Ans. $N - n = n[M/m - 1]$.

3.10–3 Modify Example 3.10-4 for the factors 2, 3, 5, 7.

3.10–4 Five balls are put into 3 boxes. What is the probability that at least one will be empty? That exactly one will be empty?

3.10–5 One marksman has an 80% chance of hitting a target, the second a 75% chance. If both shoot what is the probability of a hit? Ans. $19/20 = 0.951....$

3.10–6 A product goes through three stages of manufacture. If the first stage has a yield of 80%, the second of 90%, and the third of 95%, what is the final yield if the defects are not found until the end? Ans. 68.4%.

3.10–7   If you use three proof readers on a text then you have seven numbers, the individual numbers found, the pairwise numbers, and the total common number, and you have to estimate only four numbers, the individual probabilities and the total number of errors. Discuss a practical approach to the problem.

3.10–8   Discuss the difference, if any, between inspection with "rejection if defective" at each stage of inspection in Exercise 3.10-6, and letting the inspection go until the end.

3.11 Summary

In this chapter we have introduced a number of methods for solving probability problems, as well as a few other intellectual tools. We have also shown that the classic concept of a fair game, though widely used, is misleading when applied to money. Finally, we have further examined the topic of robustness of the solution of probability problems by supposing that the initial probabilities are not exactly known but have small errors $e(i)$. Thus for the original uniform probability assignment where we originally assumed the individual probabilities were, say, $1/n$, we must now consider the change of the solution where we now have probabilites $1/n + e(i)$ (for small $e(i)$ of course). It is seldom that anything is absolutely uniform in nature, and hence the robustness of the solution is an important aspect of the problem if we are to act on the computed results.

4

Countably Infinite Sample Spaces

4.1 Introduction

Among the aims of this book are: (1) to be clear about what is being assumed in the various probability models, (2) when possible to use regular methods and shun cute solutions to isolated problems, (3) to develop systematically the reader's intuition by choice of problems examined and the analysis of why the results are as they are, and finally, (4) show how to make reasonableness checks in many cases.

In the preceding Chapters we examined finite sample spaces, and they bring us naturally to countably infinite sample spaces when we ask questions such as: "If you repeatedly toss a coin when will the first head occur?" There is no finite number of tosses within which you are certain (probability = 1) that you will get a head, hence you face the countably infinite sample space H, $TH, TTH, TTTH, TTTTH, \ldots$ with probabilities (p = probability of a H)

$$p, \ qp, \ q^2p, \ q^3p, \ q^4p, \ldots \tag{4.1-1}$$

for the first head on the first, second, third, fourth, ... toss.

This sample space, (4.1–1), describes *any* independent two way choice of things. Although it seems like a simple step to go from a finite to an infinite sample space we will be careful to treat, as one does in the calculus, this infinite as a *potential infinite* (limit process), and not as the actual infinite. Again, the sample space (4.1–1) describes *any* independent two way choice of things.

Mathematical Aside: *Infinite Sums*

Now that we are in infinite sample spaces we will often need various simple infinite sums which are easy to derive once you know the method. (See

Section 3.8 for other sums.) The sum

$$\sum_{k=0}^{\infty} x^k = \frac{1}{1-x} \qquad (\,|x|<1\,) \qquad\qquad (4.1\text{--}2)$$

is a simple infinite geometric progression with the common ratio $r = x$ and starting value $a = 1$. Another way to see this sum is to imagine dividing out the right hand side to get the successive quotient terms $1, x, x^2, \ldots$.

If we differentiate (4.1–2) with respect to x we get (drop the $k = 0$ term)

$$\sum_{k=1}^{\infty} kx^{k-1} = \frac{1}{(1-x)^2} \qquad\qquad (4.1\text{--}3)$$

A more useful form is obtained by multiplying through by x to get

$$\sum_{k=1}^{\infty} kx^k = \frac{x}{(1-x)^2} \qquad\qquad (4.1\text{--}4)$$

Differentiate this again to get

$$\sum_{k=1}^{\infty} k^2 x^{k-1} = \frac{1+x}{(1-x)^3} \qquad\qquad (4.1\text{--}5)$$

and multiplying by x gives

$$\sum_{k=1}^{\infty} k^2 x^k = \frac{x(1+x)}{(1-x)^3} \qquad\qquad (4.1\text{--}6)$$

similar results for higher powers can be obtained the same way.

All this can be done by: (1) begining with the finite sum

$$\sum_{k=1}^{N-1} x^k = \frac{1-x^N}{1-x} = \frac{1}{1-x} - \frac{x^N}{1-x}$$

(2) doing the corresponding operations, and then (3) taking the limit as the number of terms approaches infinity. The results are the same since when $|x| < 1$ then $\lim |x|^N \to 0$ as $N \to \infty$.

We also need to look at the corresponding generating functions and simple operations on them. We have seen (see also Section 2.7) the power of the operations of differentiation and integration on generating functions. We now examine a few less obvious operations.

Let the generating function be

$$G(t) = \sum_{n=0}^{\infty} a_n t^n$$

and assume that all a_i with negative indices are zero (it is only a matter of notational convenience). Then

$$(1-t)G(t) = \sum a_n t^n - \sum a_n t^{n+1}$$

$$= \sum (a_n - a_{n-1})t^n = \sum \Delta a_{n-1} t^n$$

(4.1–7)

Repeated application will get higher differences.

If instead of multiplying we divide by $(1-t)$ then we get

$$\frac{G(t)}{1-t} = \sum_n a_n t^n \sum_k t^k$$

$$= \sum_k \sum_n a_n t^{n+k}$$

$$= a_0 + a_1 t + a_2 t^2 + a_3 t^3 + \cdots$$

$$a_0 t + a_1 t^2 + a_2 t^3 + \cdots$$

$$+ a_0 t^2 + a_1 t^3 + \cdots$$

$$+ a_0 t^3 + \cdots$$

We now set, for notational convenience,

$$b_m = \sum_{k=0}^{m} a_k$$

(4.1–8)

where the b_m are the running cumulative partial sums of the coefficients; hence

$$\frac{G(t)}{1-t} = \sum_k b_k t^k$$

(4.1–9)

If the a_k form a probability distribution then the limit of b_k as $k \to \infty$ is 1. We can set $c_k = 1 - b_k$ (the c_k are the complements of the cumulative distribution hence are the tails of the distribution). Then from (4.1–2) the generating function is

$$\frac{1 - G(t)}{1-t} = \sum c_k t^k$$

(4.1–10)

We can get the alternate terms from a generating function by using

$$\frac{G(t) + G(-t)}{2} = \sum_k a_{2k} t^{2k}$$

If we think of $t^2 = u$ and then replace u by t we have the generating function with only the even numbered coefficients

$$\frac{G(t^{1/2}) + G(-t^{1/2})}{2} = \sum_k a_{2k} t^k \tag{4.1-11}$$

This idea may be carried further and we can select the coefficients in any arithmetic progression. For the next case, and this implies the general case, we note that above we effectively used

$$x^2 - 1 = (x - 1)(x + 1)$$

hence we examine

$$x^3 - 1 = (x - 1)(x - \omega)(x - \omega^2)$$

where $\omega = \{-1 \pm i\sqrt{3}\}/2$ is a cube root of 1, and we use the fact that the sum of the roots is 0, and the sum of products of the roots taken two at a time is also zero. It is easy to work out the details of what the sums of various powers of the cube root of 1 are and show that

$$\frac{G(t) + G(\omega t) + G(\omega^2)}{3}$$

picks out every third term of the series. Suitable modifications will get the terms shifted by any amount as

$$\frac{G(t) - G(-t)}{2}$$

picks out the odd terms only in a generating function

$$\frac{G(t^{1/2}) - G(-t^{1/2})}{2t^{1/2}} = a_1 + a_3 t + a_5 t^2 + \cdots \tag{4.1-12}$$

These devices, and others, are often useful in finding answers to special problems that may arise.

It is easily seen that if we are to substitute numbers for t in the generating function then we must be concerned with the convergence of the series. But if we are merely using the linear independence of the powers of t and equating like terms in two expansions that are known to be equal, then the question of their convergence does not arise. Their convergence is a matter of delicate arguments and there is not universal agreement on what may or may not be done in these matters.

Exercises 4.1

4.1–1   Find the generating function for the coefficients a_{4k} from the generating function of the a_k.

4.1–2   Find the generating function for the coefficients a_{3k+1}.

4.1–3   Find the generating function for the numbers k^3.

4.1–4   Show that $\lim_{t\to 1}\{1 - G(t)\}/\{1 - t\}$ approaches the right limit.

4.1–5   Find the generating function for the coefficients a_k/k.

4.1–6   Find the generating function for the coefficients a_k/k^2.

4.1–7   Find the generating function for $1/en!$. Show that $G(1) = 1$.

4.1–8   Find the generating function for the kth differences of the coefficients.

4.1–9   State how to decrease the subscripts of the coefficients of a generating function. How to increase them.

4.1–10   Find $\sum_k k^4 t^k$.

4.1–11   Show that the generating function for the coefficients $x^{n+1}/(n+1)$ is a log.

4.1–12   Find an integral representation for the series $\sum_1^\infty 1/n^2$.

4.2 Bernoulli Trials

Suppose we have a situation, like the coin toss, good or defective component, life or death, etc. where there is only one of two possible outcomes. We also suppose that the trials are independent and of constant probability. We ask when the event occurs for the first time.

Example 4.2–1   *First Occurrence*

Let the probability of the event be $p \neq 0$, and let X be the random variable that describes the first occurrence of the event. Then we have (repeating 4.4–1)

$$\Pr\{X = 1\} = p$$

$$\Pr\{X = 2\} = qp$$

$$\Pr\{X = 3\} = q^2 p$$

$$\cdots$$

$$\Pr\{X = k\} = q^{k-1}p$$

$$\cdots$$

These are the same probabilities as those of the sample space (4.1–1), and they form a *geometric probability distribution*.

Have we counted all the probability? We compute the infinite sum (via the usual calculus approach of first doing in our head the finite sum and then examining the result as the number of terms approaches infinity). We have the standard result (using 4.1–2)

$$\sum_{k=1}^{\infty} \Pr\{X = k\} = \sum_{k=1}^{\infty} pq^{k-1}$$

$$= \frac{p}{1 - q} = \frac{p}{p} = 1$$

Thus the sample space has a total probability of 1 as it should.

This shows us how to get the generating function of the distribution. The variable t is a dummy variable having no real numerical value, but since we will later set it equal to 1 we must have the convergence of the infinite geometric progression for all $q < 1$ (at $q = 1$ the event cannot occur). The generating function is, therefore, from (4.1–2),

$$G(t) = \sum_{k=1}^{\infty} pq^{k-1}t^k = \frac{pt}{1 - qt} \tag{4.2-2}$$

To get the mean and variance we need the first two derivatives evaluated at $t = 1$, (see equations 2.7–3 and 2.7–4)

$$G'(t) = \frac{(1 - qt)p - pt(-q)}{(1 - qt)^2} = \frac{p}{(1 - qt)^2}$$

$$G'(1) = p/p^2 = 1/p$$

$$G''(t) = \frac{2pq}{(1 - qt)^3} \tag{4.2-2}$$

$$G''(1) = 2q/p^2$$

Hence the mean (expected time) is

$$E\{X\} = 1/p \tag{4.2-3}*$$

In words, we have the important result that *the expected value of a geometric distribution is exactly the reciprocal of the probability p.* Since

$$E\{X^2\} = G''(1) + G'(1) = 2q/p^2 + 1/p$$

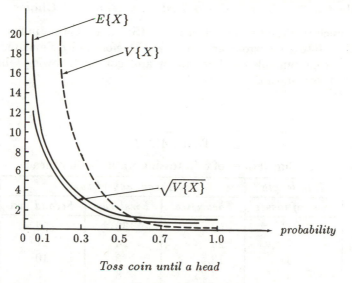

Toss coin until a head

FIGURE 4.2–1

the variance is, from 2.7–4,

$$V\{X\} = E\{X^2\} - E^2\{X\} = \frac{2q}{p^2} + \frac{1}{p} - \left(\frac{1}{p}\right)^2$$

$$= \frac{2q + p - 1}{p^2} = \frac{q}{p^2}$$

(4.2–4)*

If, as for a well balanced coin, $p = 1/2$ then the expected time to a head is two tosses, and the variance is also 2. In Figure (4.2–1) we plot as a function of p the mean, variance, and the square root of the variance, which is often used as a measure of what to expect as a reasonable deviation from the mean.

These results, (4.2–3) and (4.2–4), are important because they allow us to solve some problems very easily, as we will see in Examples (4.2–3) and (4.2–5).

Example 4.2–2   *Monte Carlo Test of a Binomial Choice*

There is much truth in the remark of Peirce (Section 2.1) that in human hands probability often gives wrong answers. We therefore twice simulated 100 trial runs on a programmable hand calculator and got the following table for the number of tosses until the first head.

TABLE 4.2–1

100 Simulations of coin tossing until the first head.

Run length	Frequency of run		
No. of tosses	Theoretical	First run	Second run
1	50	53	52
2	25	24	27
3	12.5	13	10
4	6.25	8	5
5	3.125	1	3
6	1.5625	1	1
7	0.7813	0	1
8	0.3906	0	0
9	0.1953	0	0
10	0.0977	0	0
11	0.0488	0	0
12	0.0244	0	1

The computed $E\{X\}$, $E\{X^2\}$ and Variance are:

	$E\{X\}$	$E\{X^2\}$	$V\{X\}$
Theoretical:	2	6	4.00
Experiment #1:	1.83	4.55	3.03
Experiment #2:	1.895	6.34	4.64

We conclude that we are not wrong by the usual factor of 2, but must be fairly close. Either longer runs, or simulations for other than $p = 1/2$, would give further confidence in the results derived. In the first run the results were too bunched, and the large variance in the second experiment arose from the single outlier of a run of length 12. You have to expect that the unusual will occur with its required frequency *if* the theory reasonably models reality. The extreme of $p = 1$ gives $E\{X\} = 1$, $V\{X\} = 0$, while from Figure 4.2-1 the other extreme $p = 0$ gives infinity to both values, and these results agree with our intuition.

Example 4.2–3   *All Six Faces of a Die*

What is the expected number of rolls of a die until we will have seen all six faces (assuming, of course, a uniform distribution and independence of the rolls)?

Using the method of random variables we define:

$X_1 = 1$ the number of rolls until we see the first new face

X_i = the number of rolls after the $(i-1)$st face until the ith face, $(i = 2, \ldots, 6)$. Hence the random variable for the total number of rolls until we see all six is simply

$$X = X_1 + X_2 + \cdots + X_6$$

and the expected number is

$$E\{X\} = \sum_{i=1}^{6} E\{X_i\}$$

But from the previous result, (4.2–3), the expected number is the reciprocal of the probability, and we have

$$E\{X_i\} = 1/\{(7-i)/6\} = 6/(7-i) \qquad (1 = 1, \ldots 6)$$

Hence we have (recall the harmonic sums from Appendix 1.A)

$$E\{X\} = 6\left\{ \frac{1}{6} + \frac{1}{5} + \frac{1}{4} + \frac{1}{3} + \frac{1}{2} + 1 \right\}$$

$$= 6H(6) = 6\left(\frac{147}{60} \right) \qquad (4.2-4)$$

$$= 14.7$$

In other words, the expected number of rolls until you see all the faces is above 14 and less than 15.

One is inclined to interpret this expected value of 14.7 rolls to mean that the probabilty of success on 14 rolls is less than 1/2 and that for 15 it is greater than 1/2. To do this is to confuse the expected value with the median value—a common error! To see this point we give a simple example. Let the random variable X have

$$P\{X = 1\} = 2/3$$

$$P\{X = 2\} = 0$$

$$P\{X = 3\} = 1/3$$

Then the expected value of X is

$$E\{X\} = 1\left(\frac{2}{3} \right) + 2(0) + 3\left(\frac{1}{3} \right) = \frac{5}{3}$$

Thus the expected number of trials is 5/3 but 2/3rds of the time you win on the first trial! Other examples are easily made up to show that the expected value and the median value can be quite different.

How about the variance? The random variables X_i were carefully chosen to be independent of each other, hence their variances add, and we have, from (4.2–4) and $q_i = 1 - p_i$,

$$\text{Var}\{X\} = \sum_{i=1}^{6} \frac{q_i}{p_i^2} = \sum_{i=1}^{6} \left[\frac{1}{p_i^2} - \frac{1}{p_i} \right]$$

The $p_i = (7 - i)/6$, and when you set $k = 7 - i$, $(k = 1, \ldots, 6)$, $p_i = k/6$ you have

$$\text{Var}\{X\} = \sum_{k=1}^{6} \left[\frac{36}{k^2} - \frac{6}{k} \right]$$

$$= 36[1 + 1/4 + 1/9 + 1/16 + 1/25 + 1/36] - 14.7 \qquad (4.2\text{–}5)$$

$$= [36 + 9 + 4 + 9/4 + 36/25 + 1] - 14.7$$

$$= 53 + 69/100 - 14.7 = 38.99$$

Thus the root mean square deviation from the expected value is 6.244..., and we therefore might expect a range for the trials running from 8.5 to 20.9, though, of course, we must expect to see many trials fall ouside these limits. The lowest possible number of trials is 6 which, from Example 3.2–1, has a probability of $6!/6^6 = 0.015432\ldots \sim 1\frac{1}{2}\%$

Example 4.2–4   *Monte Carlo Estimate*

To check both the particular case, and the model generally, we did a Monte Carlo simulation using 1000 trials. Part of the results are given in Table 4.2–2.

TABLE 4.2–2

Monte Carlo Until All Six Faces of a Die Appear.

Number of trials n	Probability cumulated p
⋮	
10	0.272
11	0.356
12	0.438
13	0.514
14	0.583
15	0.644
16	0.698
17	0.745
18	0.785
19	0.819
20	0.848
⋮	

and we see that by linear interpolation for $p = 1/2$ we have 12.815 as the median number of trials. This is somewhat far from the average number 14.7 because the distribution is skewed a bit. It is important to stress that the median and average can be quite different—the median is the item in the middle and is relatively little affected by the tails of a distribution, while the average covers all the distribution and can be strongly affected by the tails.

Example 4.2–5   *Card Collecting*

Often packages which you buy in a store include a card with a picture, or other items of a set, and you try to collect all of the N possible cards. What is the expected number of trials you must make?

This is a simple *generalization* of the six faces of a die, (Example 4.2–3). The expected value is now

$$E\{X\} = N\left[\frac{1}{N} + \frac{1}{N-1} + \frac{1}{N-2} + \cdots + 1\right]$$

$$= NH(N)$$

(4.2–6)

where $H(N) = 1 + 1/2 + 1/3 + \cdots + 1/N \sim ln\ N$, the truncated harmonic series (see Appendix 1.A). The variance is correspondingly

$$\text{Var}\{X\} = N^2 \sum_{k=1}^{N} \frac{1}{k^2} - \left\{ N \sum_{k=1}^{N} \frac{1}{k} \right\}^2$$

These sums are easily bounded, as shown in Appendix 4.B, by

$$ln\ N + \frac{N+1}{2N} \geq \sum \frac{1}{k} \geq ln\ N + \frac{N+1}{2N} - \frac{\left(1 - \frac{1}{N^2}\right)}{8}$$

$$\frac{3}{2} - \frac{1}{N} + \frac{1}{2N^2} \geq \sum \frac{1}{k^2} \geq \frac{3}{2} - \frac{1}{N} + \frac{1}{2N^2} - \frac{\left(1 - \frac{1}{N^3}\right)}{4}$$

so the general case is easily estimated. The differences between the bounds are the terms on the right extremes of the equations and dropping the $1/N^2$ terms gives the values $1/8$ and $1/4$ respectively as the differences between the lower and upper bounds. Approximations may be closer many times.

Example 4.2–6   *Invariance Principle of Geometric Distributions*

The probability of not getting a head in n independent trials is q^n which we shall label as

$$P(n) = q^n$$

This is a geometric progression, and has the following property:

$$P(n) = P(m)P(n - m) \qquad \text{for all } m \text{ between 1 and } n. \tag{4.2–7}$$

But this is the same, using conditional probabilities (Section 1.8), as

$$P(n) = P(m)P(n - m \,|\, m)$$

In words, we have that the probability of going n steps without the event happening is the same as the probability of going $m < n$ steps multiplied by the probability of going $n - m$ steps conditioned on having gone m steps. Thus, comparing the two equations we see that having gone m steps does not affect in any way the probability of continuing—that for the case of a geometric progression the past and future are disconnected. In different words, the probability of getting a head is independent of how many tails you have thrown—but that is exactly the concept of independence! There is an invariance in a geometric probability distribution—if you get this far without the event happening then the expected future is the same as it was before you started!

The converse is also true; if you have this property then the discrete sequence must be simply the powers of the probability of a single event as can be seen by going step by step.

Example 4.2–7   *First and Second Failures*

Suppose we have a Bernoulli two way outcome, the toss of a coin, the test of success or failure, of life or death, of good or bad component, with probability p of a failure (which is the event we are looking for), and hence q of success (not failing). The generating function for the first failure is given by (4.2–2)

$$G(t) = \frac{pt}{(1 - qt)} \qquad G(1) = 1$$

$$G'(t) = \frac{p}{(1 - qt)^2} \qquad G'(1) = \frac{1}{p}$$

$$G''(t) = \frac{2pq}{(1 - qt)^3} \qquad G''(1) = \frac{2q}{p^2}$$

with the corresponding mean $1/p$ and variance q/p^2, (4.2-3) and (4.2-4). The time to second failure will be just the convolution of this distribution with itself since the time to second failure is the sum of the time to the first plus the time to the second failure. For the second failure we have, therefore, the generating function

$$H(t) = G^2(t) = \left[\frac{pt}{1 - qt} \right]^2 \qquad H(1) = 1$$

The convolution clearly sums up exactly all the possibilities; the probability that the first failure occurred at each possible time in the interval and that the second occurred exactly at that point where the sum is the coefficient of t^k. By expanding $(1 - qt)^{-2}$ using (4.1–3) we can get the actual numbers, kpq^{k-1}.

To find the mean and variance of the distribution $H(t) = G^2(t)$ we have

$$H'(t) = 2G(t)G'(t)$$

hence the mean time to second failure is, since $G'(1) = 1/p$

$$H'(1) = 2(1)(1/p) = 2/p = 1/p + 1/p$$

as you would expect. For the second derivative we have, since $G''(1) = 2q/p$,

$$H''(t) = 2G(t)G''(t) + 2[G'(t)]^2$$

$$H''(1) = 2(1) \left(\frac{2q}{p^2} \right) + 2 \left(\frac{1}{p} \right)^2 = \frac{(4q + 2)}{p^2} = \frac{6 - 4p}{p^2}$$

hence the variance is

$$V = \frac{6 - 4p}{p^2} + \frac{2}{p} - \frac{4}{p^2} = \frac{6 - 4p + 2p - 4}{p^2}$$

$$= \frac{2 - 2p}{p^2} = \frac{2q}{p^2}$$

and we see again that for independent random variables the mean and variance add. For $q = 0$, $p = 1$, we have the mean $= 2$ and variance $= 0$ as it should. For $p = 1/2$ we get mean $= 4$ and variance $= 4$. See Figure 4.2–1.

Example 4.2–8   *Number of Boys in a Family*

Suppose there is a country in which married couples have children until they have a boy, and then they quit having more children. Suppose, also, that births are a Bernoulli process. What is the average fraction of boys in a family?

First, let $p =$ the probability of a boy, hence $q = 1 - p =$ the probability of a girl. We have

Size of family	Probability
1	p
2	qp
3	$q^2 p$
...	...
k	$q^{k-1}p$
...	...

The average family size will be from (4.1–3)

$$\sum_{k=1}^{\infty} kq^{k-1}p = \frac{p}{(1-q)^2} = \frac{1}{p}$$

The ratio of boys to the total population is, since it is a Bernoulli process, exactly p. But let us check this by direct computation. We have for the following expression, in the numerator the probability of a boy in the family for each size of family and in the denominator probability of there being the corresponding number of children in the family, each arranged by the size of the family,

$$\frac{p + qp + q^2 p + \cdots}{1 + 2qp + 3q^2 p + \cdots} = \frac{\dfrac{p}{1 - q}}{\dfrac{p}{(1 - q)^2}} = p$$

We are now ready for the original question, what is the fraction of boys in the average family. We have

$$(1)p + \frac{qp}{2} + \frac{q^2 p}{3} + \cdots$$

$$= p\left[1 + \frac{q}{2} + \frac{q^2}{3} + \cdots\right]$$

(4.2–9)

To sum the series in the square bracket we label it $f(q)$. Then

$$f(q) = 1 + \frac{q}{2} + \frac{q^2}{3} + \cdots$$

$$qf(q) = q + \frac{q^2}{2} + \frac{q^3}{3} + \cdots$$

Differentiate with respect to q

$$\{qf(q)\}' = 1 + q + q^2 + \cdots = \frac{1}{1-q}$$

We integrate both sides. (Note that $0f(0) = 0$ and that $ln\,1 = 0$ so the constant of integration is 0.)

$$qf(q) = -ln(1-q) = -ln\,p$$

From this we get back to the original expression we wanted (4.2–9)

$$-\frac{p}{q}\,ln\,p$$

(4.2–10)

as the average fraction of boys in the family. For $p = 1/2$ we get

$$ln\,2 = 0.693\ldots \sim 0.7$$

Thus half of all children are boys, but the fraction of boys in the average family is almost 7/10. The difference is, of course, that we are taking different averages.

As a check on the formula (4.2–10) we ask what value we get if $p \rightarrow 0$, (boys are never born). The result is the fraction of boys approaches 0 in the families. If $p \rightarrow 1$ (only boys are born) the formula approaches the value 1, just one child in each family, and again this is what you would expect. Hence the formula passes two reasonable checks, though at first sight the value for $p = 1/2$ seems peculiar.

Example 4.2–9   *Probability that Event A Precedes Event B*

If there are three kinds of events, A, B, and C with corresponding probabilites $p(a)$, $p(b)$, and $p(c)$, what is the probability that event A will precede event B?
In the sample space of sequences of events the sequences

$$\{C^{n-1}A\}$$

will all be successes and only these. This probability is, (since $p(a) + p(b) + p(c) = 1$)

$$P(A) = p(a) \sum_0^\infty p^n(c) = \frac{p(a)}{1 - p(c)}$$

$$= \frac{p(a)}{p(a) + p(b)}$$

Similarly for the event B preceding A

$$P(B) = \frac{p(b)}{p(a) + p(b)}$$

Note that the answer does not depend on $p(c)$ and that in practice event C can be the sum of many indifferent events (with respect to events A and B).

Example 4.2–10   *Craps*

The rules for the game of "craps" are that on the first throw of two dice:

>   You win on the sum 7 or 11
>   You lose on the sum 2, 3, or 12
>   Else this throw determines your "point."

If you get a "point" you continue to throw the pair of dice until

>   You win if you throw your point
>   You lose if you throw a 7.

Using the notation $p(i) = $ probability of the sum of the two dice being i, you win with probability (using the results of Example 4.2–9)

$$P = p(7) + p(11) + \sum_{i=4,5,6,8,9,10} p(i) \frac{p(i)}{p(i) + p(7)}$$

By symmetry

$$p(2 + j) = p(12 - j) = \frac{(1 + j)}{36}, \quad i = 0, \ldots, 5$$

Hence

$$P = p(7) + p(11) + 2 \sum_{4,5,6} \frac{p^2(i)}{p(i) + p(7)}$$

Putting in the numbers we have

$$P = \frac{6}{36} + \frac{2}{36} + \frac{2}{36} \left\{ \frac{3^2}{3+6} + \frac{4^2}{4+6} + \frac{5^2}{5+6} \right\}$$

$$= \frac{1}{18} \left\{ 3 + 1 + \frac{9}{9} + \frac{16}{10} + \frac{25}{11} \right\}$$

$$= \frac{1}{18} \left\{ 5 + \frac{8}{5} + \frac{25}{11} \right\} = \frac{1}{18} \left\{ 5 + \frac{88 + 125}{55} \right\}$$

$$= \frac{1}{18} \left\{ \frac{275 + 213}{55} \right\} = \frac{1}{18} \left\{ \frac{488}{55} \right\} = \frac{1}{9} \left\{ \frac{244}{55} \right\}$$

$$= 0.492929\ldots$$

Thus the game is slightly unfavorable to the person with the dice, and this explains why the "house" does not handle the dice.

Exercises 4.2

4.2–1 Find the exact coefficients for Example 4.2–7 by long division.

4.2–2 Estimate how long you can expect to go to get all of a set of 100 items.

4.2–3 Show that the time to third failure is the cube of the corresponding generating function. Extend to the time for the kth failure.

4.2–4 Make a plot of (4.2–10) as a function of p. Discuss its meanings.

4.2–5 In the game of "odd man out" (the person who pays the bill) find the expected duration of play. Ans 4/3 tosses.

4.2–6 If n people play the game of odd man out what is the expected duration of play? Ans. $2^{n-1}/n$.

4.2–7 If an urn has w white balls, r red balls, and b black balls, and you draw repeatedly (with replacement) what is the probability of getting a white ball before a black ball?

4.2–8 With the roll of two dice what is the probability of getting a 7 before getting an 11? Ans. 3/4.

4.2–9 In drawing cards at random with replacement and shuffling before the next draw, what is the probability of getting a spade before a red card?

4.2–10 From an urn with r red balls, w white balls, and b black balls, what is the probability of drawing a white ball before a black ball?

4.2–11 In an urn with 1 black ball, 2 red balls, and w white balls, what is the probability of getting a white ball before the black ball? Ans. $w/(w+1)$.

4.2–12   If half the cars coming down a production line are white, one third are black, one sixth are green and one sixth are mixed, what is the probability that you will see a white car before a black car? Ans. 3/5.

4.3   On the Strategy to be Adopted

In view of Peirce's remark (section 2.1) about the unreliability of probability computations, even in the hands of experts, we need to consider very carefully how we are to treat, with reasonable confidence, the topic of infinite sample spaces. We must do this because probability modeling and the corresponding computations enter into many important projects in science and engineering, as well as into medicine, economics and social planning. Furthermore, the field of probability has the property, barely alluded to by Peirce, that often our intuition cannot provide the usual checks on the reliability of the results that are used in all other fields of human activity.

Peirce identified a major weakness as the lack of uniform methods (and we have frequently been guilty of this); most probability books use various trick methods to get the answers. Furthermore, the modeling is often very remote from reality and leaves the user in great doubt as to the relevance of the results obtained. Whenever we can we need to adopt elementary, uniform, easy to understand methods, and to avoid "hi falutin," "trick" or "cute" mathematics. But if we are to handle a wide class of problems we will need powerful mathematical methods. We assume the ability to handle the calculus, simple linear algebra, and ordinary differential equations. The theory of linear difference equations with constant coefficients, which closely parallels that for linear differential equations, is given briefly in Appendix 4.A. We will avoid many of the methods of advanced mathematics since they can at times give very peculiar results whose application to the real world are some times dubious and occasionally clearly not appropriate (see Section 4.8) [St].

It is apparently a fact of life that discrete problems are not as often solvable in a nice, closed form as are the corresponding continuous problems. An examination of the calculus, which is the main tool for the continuous case, shows the difference between the two fields; compare both differentiation and integration of the calculus with differencing and summation for the finite calculus. In differentiation just before the limit is taken there are usually many terms, and most of them vanish in the limit leaving a comparatively simple expression, but in the finite differences they do not vanish. For integration there are three main tools; (1) rearrangement, (2) integration by parts, and (3) change of variable. In the corresponding summation of infinite series, we have the first two but not the third—there can be only trivial changes of variable since we must preserve the equal spacing of the terms. And it is the third of these tools that is the most widely used in the calculus to do integrations. Thus we cannot hope to solve as wide a class of problems in closed form in the discrete mathematics as we can in the continuous. This is one of the reasons we will be driven to studying the topic of continuous probability distributions in Chapter 5.

If we cannot expect to get many of the solutions to our problems in a nice, useful form, what can we do? The path often taken in text books is to carefully select the problems to be solved, but this leaves much to be desired. We know from Section 2.7 that if we can get the generating function then at least we can get the mean and variance of the distribution by simple differentiation. Thus we will focus on first getting the generating function, Section 4.5, and then the complete solution, if ever, from it. The generating function contains the distribution itself, and often we can get the distribution if we will do enough work (Section 4.6). It is also true that from the generating function alone we can get many properties of the distribution beyond the moments if we do more calculation.

The method of *state diagrams* is increasingly used in diverse fields such as electrical engineering and computer science, and has proved to be very powerful. The state diagram, by its recursive nature, allows us to represent *in a finite form* the potentially infinite sample space of possible outcomes. Hence we will concentrate on this *uniform method* as our main tool (as Peirce recommended). We must also give many non-intuitive examples to build up your intuition. In Section 3.6 we solved recursive problems by implicitly using state diagrams using two states, success and failure, and giving the transition probabilities of going from one to the other on successive steps.

The finite state diagram method also has the interesting property that the paradoxes to be discussed in Section 4.8 *cannot* be represented as finite state diagrams since they require a potentially infinite storage device to represent the amount of money to be paid and furthermore violate the requirement that state diagrams we will use generally have constant transition probabilities.

4.4 State Diagrams

It is easier to show what state diagrams are by giving examples than it is to define them clearly. We begin with a simple problem, the distribution of the number of tosses of a coin until two heads appear in a row. We will proceed slowly, discussing why we do what we do at the various stages, and after this problem we will attack a slightly harder one which will indicate the general case.

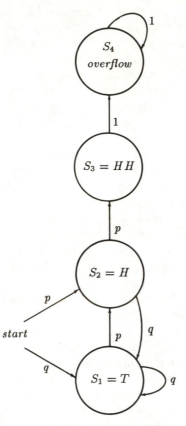

Two heads in a row

FIGURE 4.4–1

This method of state diagrams and difference equations has been already used in the problems of Section 3.5 where we set up the transition probabilities from the various states and solved the corresponding difference equations with their boundary conditions. We now turn to further Examples.

Example 4.4–1   *Two Heads in a Row*

We begin with the state diagram. The states we need are S_1 (T, tail), S_2 (H, head), S_3 (HH, two heads in a row), and S_4 (the overflow state) The last is needed if we want to apply *the conservation of probability rule* so that the sum of all the probabilites is 1. See Figure 4.4-1 where each state is shown as well as the transition probabilities. It also allows the easy calculation of the expected value of some desired state.

It is well to work on a general case rather that the special case of $p = 1/2$ for H and T so that we can follow the details more clearly. It is a curious fact that often in mathematics the general case is easier to handle than is the special case! Therefore, we suppose that on the first toss the probability of H is p, and the probability of a T is q, shown in the figure on the left by arrows coming into the state diagram.

We now write the *transition equations* for the state $S_i(n)$ in going from trial $(n-1)$ to trial n, (see Examples in Section 3.6 for similar equations). Let $S_i(n)$ be the probability of being in state S_i at time $n \geq 2$. Then for state S_1 (T, tails) we have

$$S_1(n) = qS_1(n-1) + qS_2(n-1)$$

For state S_2 (H, heads) we have

$$S_2(n) = pS_1(n-1)$$

For state S_3 (HH, two heads in a row) we have

$$S_3(n) = pS_2(n-1)$$

And for state S_4 (the overflow state) we have

$$S_4(n) = S_3(n-1) + S_4(n-1)$$

If we add all these equations we get

$$S_1(n) + S_2(n) + S_3(n) + S_4(n) = (q+p)S_1(n-1) + (q+p)S_2(n-1)$$
$$+ S_3(n-1) + S_4(n-1)$$

and we see that probability is conserved *provided* we use the overflow state $S_4(n)$ which has no influence on the problem. We now drop the state $S_4(n)$, though in a difficult problem we might include it to get a further consistency check on the results.

We now have a set of three difference equations,

$$S_1(n) = qS_1(n-1) + qS_2(n-1)$$

$$S_2(n) = pS_1(n-1) \qquad\qquad (4.4\text{--}1)$$

$$S_3(n) = pS_2(n-1)$$

together with the initial conditions, $S_1(1) = q$, $S_2(1) = p$, $S_3(1) = 0$. Note that the starting total probability is 1.

There are various methods for solving these equations. The first, and somewhat inelegant, method we will use to solve the equations 4.4–1 is to eliminate the states we are not interested in, $S_1(n)$ and $S_2(n)$. When we eliminate $S_2(n)$ (actually $S_2(n-1)$) we get the equations

$$S_1(n) = qS_1(n-1) + qpS_1(n-2)$$

$$S_3(n) = p^2 S_1(n-2)$$

We next eliminate state $S_1(n)$ by substituting the $S_1(n)$ from the second equation (with the argument shifted by 2 and by 1) into the first,

$$S_3(n+2)/p^2 = (q/p^2)S_3(n+1) + (q/p)S_3(n)$$

We easily see that the initial conditions for state $S_3(3)$ are

$$S_3(1) = 0, \quad S_3(2) = p^2$$

Multiply the difference equation through by p^2 to get it in the standard form

$$S_3(n+2) - qS_3(n+1) - pqS_3(n) = 0,$$
$$S_3(1) = 0, \quad S_3(2) = p^2 \tag{4.4-2}$$

To check this equation we try the two extreme cases, $p = 0$ which gives $S_3(n) = 0$ for all n, and the case $p = 1$ which gives $q = 0$ and $S_3(n) = 1$ for $n = 2$ and 0 otherwise.

Equation (4.4-2) is a homogeneous second order linear difference equation with constant coefficients, and according to Appendix 4.A it generally has eigenfunctions of the form r^n. Substituting r^n into the difference equation (4.4-2) and dividing out the r^n we get the *characteristic equation*

$$r^2 - qr - qp = 0 \tag{4.4-3}$$

whose roots are

$$r_1 = (1/2)[q + \sqrt{q^2 + 4pq}] = (q + R)/2$$
$$r_2 = (1/2)[q - \sqrt{q^2 + 4pq}] = (q - R)/2 \tag{4.4-4}$$

where the quadratic irrationality is contained in

$$R = \sqrt{q^2 + 4pq}$$

Note also the relationships, $r_1 + r_2 = q$, $r_1 r_2 = -pq$, $r_1 - r_2 = R$ which we will need later. How big is R, or rather R^2? We set

$$R^2 = f(p) = q^2 + 4pq$$
$$= 1 + 2p - 3p^2 = (1-p)(1+3p) \tag{4.4-5}$$

The maximum of the curve $f(p)$ occurs at $f'(p) = 0$, namely at

$$f'(p) = 2 - 6p = 0, \qquad \text{or } p = 1/3.$$

At this maximum place

$$f(1/3) = (1 - 1/3)(1 + 1) = 4/3$$

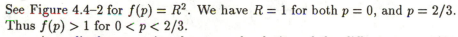

$$R^2 = 1 + 2p - 3p^2 = (1 - p)(1 + 3p)$$

Radical

FIGURE 4.4–2

See Figure 4.4–2 for $f(p) = R^2$. We have $R = 1$ for both $p = 0$, and $p = 2/3$. Thus $f(p) > 1$ for $0 < p < 2/3$.

Accordingly, we write the general solution of the difference equation (4.4–2) as ($p \neq 1$)

$$S_3(n) = C_1 r_1^n + C_2 r_2^n \tag{4.4–6}$$

where C_1 and C_2 are constants to be determined by fitting the initial conditions $S_3(1) = 0$, $S_3(2) = p^2$. Thus we get

$$S_3(1) = 0 = C_1 r_1 + C_2 r_2$$

$$S_3(2) = p^2 = C_1 r_1^2 + C_2 r_2^2$$

To solve for C_2 multiply the top equation by r_1 and subtract the lower equation

$$-p^2 = C_2[r_2(r_1 - r_2)] = C_2 r_2 R \Longrightarrow C_2 = -p^2/r_2 R$$

To solve for C_1 multiply the top equation by $-r_2$ and add to lower equation

$$p^2 = C_1[r_1(r_1 - r_2)] = C_1 r_1 R \Longrightarrow C_1 = p^2/r_1 R$$

Hence the solution to the difference equation satisfying the initial conditions is

$$S_3(n) = \frac{p^2}{R}[r_1^{n-1} - r_2^{n-1}]$$

$$= \left(\frac{p^2}{R}\right)\left[\left\{\frac{p+R}{2}\right\}^{n-1} - \left\{\frac{p-R}{2}\right\}^{n-1}\right] \tag{4.4–7}$$

If we examine 4.4–7 closely, we see that the radical R will indeed cancel out and we will have a rational expression in p and q. We can check it for $n = 1$, and $n = 2$, and of course it fits the initial conditions since we chose the constants so that it would. If we use $n = 3$ we get

$$S_3(3) = \left\{ \frac{p^2}{r_1 - r_2} \right\} [r_1^2 - r_2^2] = p^2[r_1 + r_2] = p^2 q$$

This is correct since to get a success in three steps you must have first a q and then two p's.

To understand this formula in other ways let us rearrange it, remembering that $r_1 - r_2 = R$,

$$S_3(n) = \left(\frac{p^2}{r_1 - r_2} \right) r_1^{n-1} \left[1 - \left(\frac{r_2}{r_1} \right)^{n-1} \right]$$

hence as n gets large only the first part in the square brackets is left and $S_3(n)$ approaches

$$\left(\frac{p^2}{r_1 - r_2} \right) r_1^{n-1} = \frac{p^2}{R} r_1^{n-1} \tag{4.4-8}$$

The approach is surprisingly rapid when the roots are significantly different (R is not small).

The cancellation of the quadratic irrationality R in this problem is exactly the same as in the classic problem of the Fibonacci numbers which are defined by

$$f_{n+1} = f_n + f_{n-1}, \qquad \text{where} \qquad f_0 = 0, \ f_1 = 1$$

The solution is well known to be (see Appendix 4.A)

$$f_n = \frac{1}{\sqrt{5}} \left[\left\{ \frac{1 + \sqrt{5}}{2} \right\}^{n-1} - \left\{ \frac{1 - \sqrt{5}}{2} \right\}^{n-1} \right] \tag{4.4-9}$$

and the radical $\sqrt{5}$ cancels out as it must since clearly the Fibonacci numbers are all integers.

Example 4.4–2   *Three Heads in a Row*

The probability of tossing three heads in a row leads again to an infinite sample space. This time we use the states:

$$
\begin{aligned}
S_0 &: \quad \emptyset \\
S_1 &: \quad T \\
S_2 &: \quad H \\
S_3 &: \quad HH \\
S_4 &: \quad HHH
\end{aligned}
$$

For a well balanced coin the probabilities of going from one state to another are as indicated in the graph of Figure 4.4–3. We see that out of every state the total probability of the paths is 1, except from state S_4. This

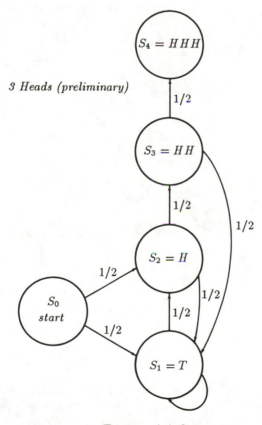

3 Heads (preliminary)

FIGURE 4.4–3

fact comes close to a rule of conservation of probability. To make the rule inviolate (and conservation rules have proved to be extremely useful in physics

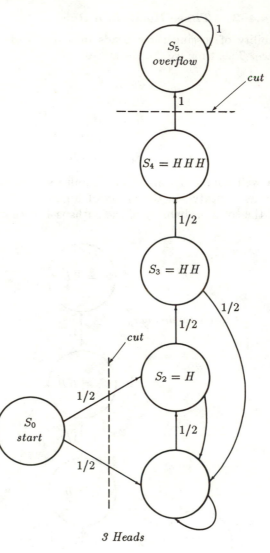

3 Heads

FIGURE 4.4–4

and other fields) we again append a further state, S_5, Figure 4.4–4 from which state it goes to itself with probability 1. As before you can think of state S_5 as the overflow state that is needed to *conserve probability*. Thus we will have the powerful check at all stages that the total probability is 1.

Using the modified diagram with state S_5 included, Figure 4.4–4, we write the equations for the probability of being in state S_i at time n as $S_i(n)$,

$i = 1, 2, \ldots, 5$. We get for the states

$$S_1(n) = (1/2)S_1(n-1) + (1/2)S_2(n-1) + (1/2)S_3(n-1)$$
$$S_2(n) = (1/2)S_1(n-1)$$
$$S_3(n) = (1/2)S_2(n-1) \qquad\qquad (4.4\text{--}10)$$
$$S_4(n) = (1/2)S_3(n-1)$$
$$S_5(n) = S_4(n-1) + S_5(n-1)$$

We notice that the equation for $S_0(n)$ does not occur since there are no entry arrows, only exit arrows. We have, if you wish, $S_0(0) = 1$, and $S_0(n) = 0$ for $n > 0$, but this state has no use in the rest of the problem.

We check for *the conservation of probability* by adding all the equations.

$$S_1(n) + S_2(n) + S_3(n) + S_4(n) + S_5(n)$$
$$= S_1(n-1) + S_2(n-1) + S_3(n-1) + S_4(n-1) + S_5(n-1)$$

Yes, the probability is conserved and the total probabilty is 1 since it checks for $n = 0$ (if we include $S_0(0) = 1$) and for $n = 1$ we have $S_1(1) + S_2(1) = 1$ with all other $S_i = 0$.

The conservation of probability allows us to think of the probability as an incompressible fluid, like water. The fluid flows around the system, and does not escape (provided we put in the overflow state, S_5 in this case); we look only at discrete times $t = 0, 1, 2, \ldots$. The fluid analogy provides a basis for some intuition as to how the system will evolve and what is reasonable to expect; it provides us with a valuable check on our results.

Equations (4.4–10) are homogeneous linear difference equations with constant coefficients. For such equations the *eigenfunctions* (see Appendix 4.A) are simply

$$r^n$$

This time instead of eliminating all states except the desired one, as we did in Example 4.4–1, we will solve them as a system. Thus we assume the representation

$$S_1(n) = A_1 r^n$$
$$S_2(n) = A_2 r^n$$
$$S_3(n) = A_3 r^n$$
$$S_4(n) = A_4 r^n$$
$$S_5(n) = A_5 r^n$$

where the A_i are constants and r will later be found to be a characteristic root. We get the system of homogeneous linear algebraic equations

$$A_1 r = (1/2)A_1 + (1/2)A_2 + (1/2)A_3$$
$$A_2 r = (1/2)A_1$$
$$A_3 r = \qquad\quad (1/2)A_2$$
$$A_4 r = \qquad\qquad\qquad (1/2)A_3$$
$$A_5 r = \qquad\qquad\qquad\qquad\qquad A_4 + A_5$$

Since we are interested in state S_4 we will apply an obvious rule that if we can "cut" a state diagram into two or more parts such that each part has only outgoing arrows that can never lead back, then for states within the part we will need to use only the eigenvalues that arise from those equations in the subset. The arrows entering the region provide the "forcing term" of the corresponding difference equations; the outgoing arrows have no further effects on the set of equations. We can therefore neglect them, for all other states; for example the state S_5 with its eigenvalue $r = 1$ and the state S_0 with its eigenvalue 0.

A standard theorem of linear algebra states that the homogeneous linear equations that remain can have a nontrivial solution if and only if the determinant of the system is 0. The determinant (after transposing terms to the left) is

$$\begin{vmatrix} r - 1/2 & -1/2 & -1/2 \\ -1/2 & r & 0 \\ 0 & -1/2 & r \end{vmatrix} = 0$$

$$(4.4\text{--}11)$$

The characteristic equation is, upon expanding the determinant,

$$r^3 - (1/2)r^2 - (1/4)r - 1/8 = 0 \qquad (4.4\text{--}12)$$

We will label the three roots of the cubic $r_1 = 0.9196\ldots$, r_2, $r_3 = -0.2098 \pm 0.3031i$.

From the linearity of these equations we know that the probability of any of these states S_i can be written as a linear combination of powers of these three eigenvalues (characteristic roots). Since there are no multiple roots we have, therefore,

$$S_4(n) = C_1 r_1^n + C_2 r_2^n + C_3 r_3^n = \sum_{1=1}^{3} C_i r_i^n \qquad (4.4\text{--}13)$$

The unknown coefficients C_i are determined by the initial conditions, but rather than finding them now we will find the generating function $G(t)$

for the probabilities of the state of $S_4(n)$. It is found by multiplying the nth of the equations (4.4–13) by t^n and summing

$$G(t) = \sum S_4(n)t^n = \sum_{i=1}^{3} \sum_{n=1}^{\infty} C_i(r_i t)^n$$

These are three infinite geometric progressions with the first term $r_i t$ and the common ratio $r_i t$, hence they are each easily summed (see 4.2–1) to get

$$G(t) = \sum_{i=1}^{3} C_i r_i t / (1 - r_i t)$$

$$= t \left[\frac{C_1 r_1}{1 - r_1 t} + \frac{C_2 r_2}{1 - r_2 t} + \frac{C_3 r_3}{1 - r_3 t} \right]$$

(4.4–14)

The common denominator of these three terms is

$$(1 - r_1 t)(1 - r_2 t)(1 - r_3 t) = D(t) \tag{4.4–15}$$

This denominator resembles the original characteristic polynomial *except* that the coefficients are reversed. This reversal comes about because when you examine a typical factor in the form

$$(1 - r_i t) = t \left[\frac{1}{t} - r_i \right]$$

you see the reversal, and hence in the product of all the factors you have t^n times the characteristic polynomial (4.4–12) in $1/t$. This, in turn, means that you merely have the polynomial with the coefficients in reverse order!

In this method the characteristic equation begins with r^n with unit coefficient. But in general it might not be exactly the polynomial since the roots determine a polynomial only to within a multiplicative constant. Any scale factor that may be accidentally included in the denominator will be exactly compensated for when we find the numerator from the initial conditions so this delicate point is of no practical significance. Thus we have the denominator of the generating function

$$D(t) = [1 - (1/2)t - (1/4)t^2 - (1/8)t^3]$$

An examination of the numerator, excluding the front factor t, shows that it is of degree 2. Thus this numerator of the generating function must be of the form,

$$N(t) = a_0 + a_1 t + a_2 t^2$$

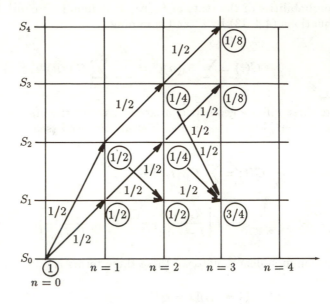

Initial conditions

FIGURE 4.4–5

where now

$$G(t) = tN(t)/D(t)$$

The coefficients a_i $(i = 0, 1, 2)$ of the numerator are determined by the initial conditions, and hence we need three of them. Thus we evolve the state diagram for $n = 1, 2, 3$ as shown in Figure 4.4-5. The figure is easy to construct. To move from left to right we take the probability at each node, multiply it by the probability of going in each direction it can, and add all the contributions at the new node.

We are interested in state $S_4(n)$ which has the initial conditions

$$S_4(0) = 0, \quad S_4(1) = 0, \quad S_4(2) = 0, \quad S_4(3) = 1/8$$

We have merely to divide the numerator of the generating function by the denominator to determine the unknown coefficients in the numerator. In this particular case we see immediately that the terms a_0 and a_1 are both zero and the next coefficient a_2 in $N(t)$ is 1/8. The generating function is, therefore,

$$G(t) = \frac{tN(t)}{D(t)} = \frac{\dfrac{t^3}{8}}{1 - \dfrac{t}{2} - \dfrac{t^2}{4} - \dfrac{t^3}{8}}$$

In this form of the solution we have neatly *avoided all of the irrationalities and/or complex numbers that may arise in solving for the roots of the characteristic equation* which in this case is a cubic.

You can see (using the water analogy) that all the probability in this problem will end up in the overflow state, hence it must all pass through the state S_4. This means, as you can easily check, that $S_4(t)$ is a probability distribution and $G(1) = 1$. Hence we can apply our usual methods. Had $G(1) \neq 1$, we could still handle the distribution as a conditional one, conditional on ending up in that overflow state, and this would require the modification of the method by dividing by the corresponding conditional probability.

To get the mean and variance we will need the first and second derivatives of $N(t)$ and $D(t)$ evaluated at $t = 1$. We have

$$
\begin{aligned}
N(t) &= t^2/8 & N(1) &= 1/8 \\
N'(t) &= 2t/8 & N'(1) &= 2/8 \\
N''(t) &= 2/8 & N''(1) &= 2/8 \\
D(t) &= 1 - t/2 - t^2/4 - t^3/8 & D(1) &= 1/8 \\
D'(t) &= -1/2 - t/2 - 3t^2/8 & D'(1) &= -11/8 \\
D''(t) &= -1/2 - 6t/8 & D''(1) &= -10/8
\end{aligned}
$$

We now have the pieces to compute the derivatives

$$
\begin{aligned}
G'(1) &= \frac{DN + DtN' - tND'}{D^2} \\
&= \frac{\left(\frac{1}{8}\right)\left(\frac{1}{8}\right) + \left(\frac{1}{8}\right)\left(\frac{2}{8}\right) - \left(\frac{1}{8}\right)\left(\frac{-11}{8}\right)}{8^2} \qquad (4.4\text{--}16) \\
&= 1 + 2 + 11 = 14
\end{aligned}
$$

After some tedious work we have

$$
G''(1) = 324 \qquad (4.4\text{--}17)
$$

Hence the mean is 14 and the variance is

$$
\text{Var} = G''(1) + G'(1) - [G'(1)]^2 = 142 \sim (12)^2
$$

Notice that the generating function of each state S_1, S_2, S_3, or S_4, is determined by its own initial conditions since they are what determines the coefficients in the numerator; *the denominator is always the same*.

Exercises 4.4

4.4–1 Derive the solution of the Fibonacci numbers.

4.4–2 Solve Example 4.4–1 in the two extreme cases $p = 1$ and $p = 0$.

4.4–3 Repeat the Example 4.4–4 but now using the matrix method.

4.4–4 In Example 4.4–1 suppose you are to end on the state TH; find the mean and variance.

4.4–5 Do Example 4.4–2 *except* using the pattern HTH.

4.4–6 What is the generating function of the probability distribution of the number of rolls until you get two 6 faces in a row?

4.4–7 Find the generating function for the probability of drawing with replacement two spades on succesive draws.

4.4–8 What is the expected value and the variance for drawing (with replacement) two aces in a row from a deck of cards?

4.4–9 In Exercise 4.4–6 what is the mean and variance of the distribution?

4.5   Generating Functions of State Diagrams

We see that the essence of the method is to draw the finite state diagram, along with the constant probabilities of transition from one state to another. The method requires:

1. A finite number of states in the diagram (unless you are prepared to solve an infinite system of linear equations and act on the results).

2. The probabilities of leaving each state are constant and independent of the past (unless you can solve systems of difference equations with variable coefficients). Note that we have handled the apparent past history by creating states, such as the state HH, which incorporate some of the past history so that the state probabilities are independent of past history. Thus we can handle situations in which the probabilities at one stage can depend on a *finite* past history.

We will not be pedantic and state all the rules necessary for the use of the method on a particular state diagram, since the exceptions are both fairly obvious and seldom, if ever, arise in practice. When they do, the modifications are easy to make. They are also called Markov chains in some parts of the literature.

The steps in getting the rational generating function are:

1. Write the difference equations for each state and set up the matrix, and then the determinant

$$(-1)^n \, |A - rI| \,=\, |rI - A| = 0$$

This determinant gives the characteristic equation. Note that the subset you want can often be cut out of the entire state diagram, and if so then you probably need only consider the eigenvalues that come from that subset of the entire diagram. Generally you need to exclude overflow states since their total probability is not a probability distribution, and the corresponding generating function has infinity for its value at $t = 1$.

2. To find the generating function write the denominator as the characteristic polynomial with coefficients reversed, and assume unknown coefficients for the numerator. (It really does not matter how you scale the denominator, the numerator will adjust for it.) Note that t factors out of the numerator because you started with the first state $n = 1$.

3. Project forward (evolve) the probabilites of the variable you are interested in to get enough values to determine the unknown coefficients in the numerator $N(t)$, and find them one by one. You then have the generating function

$$G(t) = tN(t)/D(t)$$

provided you start with the case $n = 1$.

4. Check that $G(1) = 1$. If it is not 1 then you have a conditional distribution (or else an error). If $G(1) = 1$ find $G'(t)$ and $G''(t)$ evaluated at $t = 1$, equations (4.4–6) and (4.4–7), and then get the mean and variance. (If these are only conditional distributions then you need to interpret them by dividing by their total conditional probabilities.)

5. If you want the detailed solution then the method of partial fractions is the next step, and is discussed in the next Section. If all you want is the asymptotic growth of the solution then this is easy to get since the solution is going to be dominated by the largest eigenvalue(s) (largest in absolute value). This can be seen when this root to the nth power is factored out of the solution, and the deviation from this dominating root easily estimated.

Example 4.5–1   *The Six Faces of a Die Again*

Example 4.4–2 solved the problem of the expected number of rolls of a single die until you have seen all six faces by what amounts to a trick method. Can we do it using state diagrams? To show that we can use the general method successfully (though it is true that often the effort is greater) we begin with the state diagram, Figure 4.5–1, with states $S_i(i = 1, \ldots, 6)$ where S_i means we have seen i distinct faces at that point. We get the equations

$$S_1(n + 1) = (1/6)S_1(n)$$

$$S_2(n + 1) = (5/6)S_1(n) + (2/6)S_2(n)$$

$$S_3(n + 1) = (4/6)S_2(n) + (3/6)S_3(n)$$

$$S_4(n + 1) = (3/6)S_3(n) + (4/6)S_4(n)$$

$$S_5(n + 1) = (2/6)S_4(n) + (5/6)S_5(n)$$

$$S_6(n + 1) = (1/6)S_5(n)$$

These equations lead to the determinant for the characteristic roots

$$\begin{vmatrix} r - 1/6 & 0 & 0 & 0 & 0 & 0 \\ -5/6 & r - 2/6 & 0 & 0 & 0 & 0 \\ 0 & -4/6 & r - 3/6 & 0 & 0 & 0 \\ 0 & 0 & -3/6 & r - 4/6 & 0 & 0 \\ 0 & 0 & 0 & -2/6 & r - 5/6 & 0 \\ 0 & 0 & 0 & 0 & -1/6 & r \end{vmatrix} = 0$$

This determinant nicely expands to the product

$$(r - 1/6)(r - 2/6)(r - 3/6)(r - 4/6)(r - 5/6)r = 0$$

The $r = 0$ root can be neglected so that the denominator becomes

$$D(t) = (1 - t/6)(1 - 2t/6)(1 - 3t/6)(1 - 4t/6)(1 - 5t/6)$$

with $D(1) = 5!/6^5$. What we want is $S_6(n)$ and we have

$$S_6(k) = 0, \quad (k = 1, \ldots, 5) \quad S_6(6) = 5!/6^5$$

Hence for the numerator we have

$$tN(t) = 5!t^6/6^5 \qquad N(1) = 5!/6^5$$

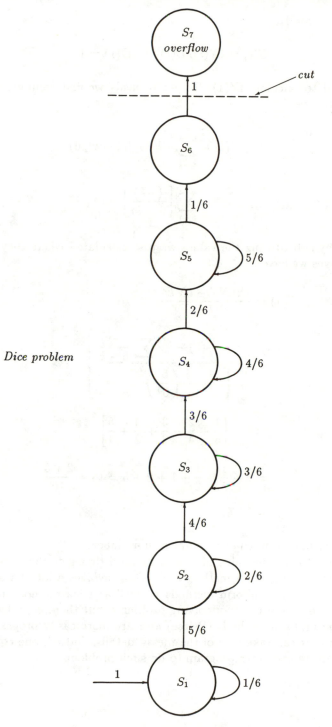

Dice problem

FIGURE 4.5–1

We have, therefore,

$$G(t) = tN(t)/D(t) \qquad G(1) = 1$$

We will need to calculate $D'(1)$. To get it easily we first compute the logarithmic derivative

$$D'(t)/D(t) = \sum_{k=1}^{5}(-k/6)/(1 - kt/6)$$

$$= \sum_{k=1}^{5}\left[\frac{(-k)}{6 - kt}\right]$$

Therefore, by substituting the values we just calculated where they occur in the derivatives we have

$$G'(1) = \frac{D(tN)' - tND'}{D^2}$$

$$= \frac{5!}{6^5}\left\{ \frac{\dfrac{5!}{6^5}(6) - \dfrac{5!}{6^5}\sum\dfrac{(-k)}{6 - k}}{\left(\dfrac{5!}{6^5}\right)^2} \right\}$$

$$= 6 + \left[\frac{1}{5} + \frac{2}{4} + \frac{3}{3} + \frac{4}{2} + \frac{5}{1}\right]$$

$$= 6 + \frac{1}{5} + \frac{1}{2} + 1 + 2 + 5 = 14 + \frac{2 + 5}{10}$$

$$= 14.7$$

which is the same answer but at a much more labor.

One has to ask the value of a trick method that gets the answer if you see the trick, vs. a general method that often involves a lot of work before you get the answer. Uniform methods as well as trick methods have their place in the arsenal of tools for solving problems, but there is probably more safety in the uniform methods (Peirce) and are more easily programmed on a computer that can take care of the messy details. Indeed, one could easily write a complete computer program to do such problems.

Exercises 4.5

4.5–1   Work the die problem if there are only three equally likely faces on the die.

4.5–2   Find the generating function for the distribution of getting three heads in a row while tossing a coin.

4.5–3   Compute the variance of the Example 4.5–1. 4.5–4 In Example 4.4–2 one sees that the generating function has the denominator (4.4–15) and that the partial fraction expansion is merely backing you up to (4.4–14), and thus the expansion in powers of t gets you to (4.4–13). Discuss the implications.

4.5–5   For Example 4.4–1 give the corresponding equations (4.4–7) and the actual terms.

4.5–6   Find the generating function (using the direct method) for getting three spades in a row from a deck of shuffled cards.

4.6   Expanding a Rational Generating Function

If you can factor the denominator into real linear and real quadratic factors (this can be done theoretically since any polynomial with real coefficients is the product of real linear and real quadratic factors, and by a computer in practice if necessary) you apply the standard method of partial fractions (remember to factor t out of the numerator before you start so that the degree of the numerator is less than that of the denominator as is required in partial fraction methods). You then divide out each factor to get it in the form of a sum of powers of t. Grouping the equal powers together you have the detailed values of the variable you are interested in.

In more detail, since the steps to do after the method of partial fractions is applied are often not discussed carefully:

(1) For a simple linear factor you have

$$1/(1 - rt) = \sum_{i=0}^{\infty} r^n t^n$$

(2) For repeated factors the derivatives of this expression with respect to r will produce the needed expansions.

(3) For a simple real quadratic $x^2 - px + q$ with complex roots we have $p^2 - 4q < 0$. We use the complex polar form for the roots

$$1 - pt + qt^2 = (1 - Ae^{ix}t)(1 - Ae^{-ix}t)$$

$$= 1 - (2A \cos x)t + A^2 t^2$$

which shows that $A = \sqrt{q}$ and, of course, $p = 2A \cos x$ which determines a real x (since $p^2 < 4q$). You now expand into partial fractions using unknown coefficients d_1 and d_2,

$$\frac{1}{(1 - Ae^{ix}t)(1 - Ae^{-ix}t)} = \frac{d_1}{1 - Ae^{ix}t} + \frac{d_2}{1 - Ae^{-ix}t}$$

After a little algebra

$$d_1 = \frac{e^{ix}}{2i \sin x} \qquad d_2 = \frac{-e^{-ix}}{2i \sin x}$$

hence

$$\frac{1}{1 - pt + qt^2} = \frac{e^{ix}}{2i \sin x} \sum_{k=0}^{\infty} A^k e^{ikx} t^k - \frac{(e^{-ix})}{(2i \sin x)} \sum_{k=0}^{\infty} A^k e^{-ix} t^k$$

$$= \sum_{k=0}^{\infty} A^k \left\{ \frac{e^{i(k+1)x} - e^{-i(k+1)x}}{2i \sin x} \right\}$$

$$= \sum_{k=0}^{\infty} A^k t^k \frac{\sin(k+1)x}{\sin x}$$

(4) For repeated quadratic factors, you differentiate this expression with respect to the parameter p to get the next higher power in the denominator, and of course make the corresponding adjustments in the numerator. To get rid of a factor of t in the numerator we have only to divide the expansion by t. With this you can add the proper combination (with and without t) to get the appropriate linear term, thus getting as a result two terms where you had one. Finally, you group the terms from the expansions for each power of t to get the coefficients of the generating function, and hence the probabilities of the state you are solving for. But remember the initial removal of the front factor of t. As before, it is the largest root in absolute value that dominates the solution for large n (or the set if there are more than 1 of that size).

To summarize, if we have the generating function we can get the mean and variance (and higher moments) by differentiation and evaluation at $t = 1$. If we want the general term of the solution then there seems to be no way other than to factor the denominator of the generating function (get the characteristic roots), and then carry out the partial fraction expansion and subsequent representation in powers of t. The coefficients of the powers of t are, from the definition of the generating function, the desired probabilities. Simple and repeated real linear and real quadratic factors can be handled in a regular fashion. But note that the algebraic (irrational) nature of the roots must cancel out in the end because the final answer is rational in the coefficients of the difference equations and the initial conditions. The well known example (discussed earlier) of this is the Fibonacci numbers (integers). However, if all we want are a few of the early coefficients then we can simply divide out the rational fraction to get the coefficients of the powers of t.

4.7 Checking the Solution

How can we further check the solution we have obtained? One way is to compute a few more values from the state diagram and compare them with those calculated from the solution. Another way of checking the solution is to generalize, or embed, the problem in a family of problems. This is often effective because the corresponding general solution contains the particular solution and may contain other, degenerate, solutions which are likely to be known to you so you can check the accuracy of the model and solution.

Example 4.7–1   *Three Heads in a Row*

Suppose we take the above problem (Example 4.4–2) except that the probability of a success is now p, and the complementary probability of failure is $q = 1 - p$. If $p = 1$ we know the solution, and if $p = 0$ we will again know the solution. If we further plot both the mean and variance as a function of p we will get a feeling for the problem and why we have, or have not, the correct result.

   We merely sketch in the steps, leaving the details for the student. The characteristic equation will be

$$r^3 - qr^2 - pqr - p^2q = 0$$

The corresponding generating function is (using the initial conditions)

$$G(t) = t\left[\frac{p^3t^2}{1 - qt - pqt^2 - p^2qt^3}\right]$$

We again adopt the notation

$$G(t) = tN(t)/D(t)$$

$$G'(t) = [D(tN)' - tND']/D^2$$

$$G''(t) = [D(tN)'' - tND'']/D^2 - 2D'G'/D$$

We have for the parts evaluated at $t = 1$

$$tN = p^3, \qquad (tN)' = 3p^3, \qquad (tN)'' = 6p^3$$

$$D = p^3, \qquad D' = 3p^3 - p^2 - p - 1, \qquad D'' = 6^3 - 4p^2 - 2p$$

$$G'(1) = [1 + p + p^2]/p^3, \qquad G''(1) = \text{(evaluate the parts above)}$$

At $p = 1/2$ this agrees with the previous results. Next we plot the values as a function of p as shown in Figure 4.7–1. If $p = 0$ then we will never get three successes in a row, while if $p = 1$ it will happen on the third trial and the corresponding variance will be zero. The curve confirms this.

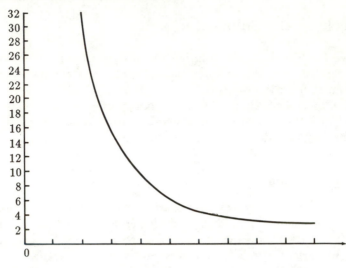

FIGURE 4.7–1

4.8 Paradoxes

The simple, countably discrete sample space we have carefully developed can give some peculiar results in practice. The classic example is the St. Petersburg paradox. To dramatize the trouble with this paradox we follow it with the Castle Point paradox. We then turn to various "explanations" of the paradoxes—there are, of course, those who feel that they need no "explanation." See also [St].

Example 4.8–1   *The St. Petersburg Paradox*

In this paradox you play a game; you toss a fair coin until you get a head, and the toss on which you *first* get the head is called the kth toss. When the first success occurs on the kth toss you are to be paid the amount

$$2^k$$

units of money and the game is over. How much should you pay to play this game, supposing you are willing to play a "fair game"?

The probability of a success on the kth toss is exactly

$$2^{-k}$$

so the expectation on the kth toss is exactly 1. Thus the sum of all the expectations is the sum of 1 taken infinitely many times! You should be willing to pay an infinite amount, or at least an arbitrarily large amount, to play.

When you think about the matter closely, and think of your feelings as you play, you will see that you would probably not be willing to play that sort of a game. For example, if you bet 128 units you will be ahead only if the first six faces are all tails—one chance in 128. Hence the question arises, "What is wrong?" You might want to appeal to the Bernoulli evaluation criterion, Sections 3.7 and 3.8, to get out of the difficulty.

Example 4.8–2   *The Castle Point Paradox*

To define this game we may take any sequence $u(n)$ of integers approaching infinity no matter how rapidly. For example, we may use any of the following sequences,

$$u(n) = n \qquad u(n) = n! \qquad \text{or} \qquad u(n) = (n!)^{n!}$$

We will pick the third for purposes of illustration. These $u(n)$ will be called "the lucky numbers." If the first head in the sequences of tosses of a fair coin occurs on a lucky number you are paid, otherwise you lose. The payment is

$$\frac{2^{u(n)}}{n}$$

and again the probability of success on this toss is

$$2^{-u(n)}$$

This time the expectation on the nth lucky number is $1/n$. But the series

$$\sum_{n=1}^{\infty} \frac{1}{n}$$

diverges to infinity, hence the total expectation is infinite.

Let us look at this game more closely. We have the table of the first 5 lucky numbers:

n	$n!$	$(n!)^{n!}$
1	1	1
2	2	4
3	6	4.6656×10^4
4	24	1.3337×10^{33}
5	120	3.1750×10^{249}

There are less than 5×10^{17} seconds in 15 billion years (the currently estimated age of the universe). There are about 3.14×10^9 seconds in 100 years so that if you tossed the coin once every microsecond (10^{-6}) you would do less than

$$3.14 \times 10^{15}$$

tosses in your lifetime, and you would not even reach the fourth lucky number. To reach the third lucky number you need 46,655 tails followed by 1 head! I could, of course, redefine the game so that the first three lucky numbers did not count, and still have the infinite series diverge to infinity and hence still have the paradox that to play the game you should be willing to pay an arbitrary amount of money.

Example 4.8–3   *Urn with Black and White Balls*

An urn contains 1 black and 1 white ball. On each drawing:

    If ball is white then end of game

    If ball is black return and add another black ball

Find the expected number of draws until you get a white ball.
The sample space of successes is

$$\{b^{n-1}w\}$$

and the total probability is

$$P = \frac{1}{2} + \frac{1}{2}\left(\frac{1}{3}\right) + \frac{1}{3}\left(\frac{1}{4}\right) + \cdots + \frac{1}{k}\left(\frac{1}{k+1}\right) + \cdots$$

$$= 1 - \frac{1}{2} + \frac{1}{2} - \frac{1}{3} + \frac{1}{3} - \frac{1}{4} + \cdots + \frac{1}{k} - \left(\frac{1}{k+1}\right)$$

$$+ \cdots$$

$$= 1$$

hence we have the total probability of getting a white ball is 1. The expected number of draws is

$$E\{K\} = \sum_{k=0}^{\infty}(k)\frac{1}{k(k+1)} = \sum_{k=0}^{\infty}\frac{1}{k+1}$$

which is the harmonic series and diverges to infinity. Hence the expected number of draws until a white ball is drawn is infinite.

Given these paradoxes some people deny that there is anything to be "explained." Hence for them there is nothing more to say.

There is, among other explanations, one which I will call "the mathematician's explanation" (meaning you have to believe in this standard kind of mathematics). The argument goes as follows: since clearly it is assumed that you have an infinite amount of money, then to enter the game the first time you simply pay me every other unit of money you have, thus both paying me the infinite entry fee and also retaining an infinite amount for the next game. On my side, I similarly pay off for finite amounts and retain an infinite amount of money. Hence after many games we both retain an infinite amount of money and the game is "fair"!

An alternate explanation uses the standard "calculus approach"; assume finite amounts and see what happens as you go towards infinity. Assume in the St. Petersburg paradox that I have a finite amount of money, say $A = 2^n$. Thus you can play at most n tosses and still be paid properly. At this point the expectation is

$$\sum_{i=1}^{n} 1 = n = \log_2 A$$

and this is the log (base 2) of the amount of money I have. This seems to be more reasonable. If the maximum you could win is, say, 2^{20} (approximately a million units) then risking 20 units is not so paradoxical. A similar analysis will apply to the Castle Point paradox. If you assume that only the first three lucky numbers are going to pay off then your expectation is

$$1 + 1/2 + 1/3 = 11/6 < 2$$

This seems, again, to be a reasonable amount to risk for the three possible payments (one of them is hardly possible). For the first two lucky numbers the expectation is $3/2$.

The confusion arises when n goes to infinity and hence both your payment and the possible winnings, A and $\log_2 A$ in the St. Petersburg paradox, both go to infinity. The two very distinct quantities, A and $\log_2 A$, both approaching infinity, are being viewed in the limit as being the same.

There are other ways of explaining the paradoxes, including either redefining the game or else using what seem to be rather artificial definitions of what a fair game is to be. We will ignore such devices.

The purpose of these paradoxes is to warn you that the careless use of even a simple countably infinite discrete probability space can give answers that you would not be willing to act on in real life. Even in this discrete model you need to stop and think of the relationship between your probability model and reality, between formal mathematics and common experience. Note again that both the St. Petersburg and Castle Point paradoxes are excluded if you adopt the finite state diagram (with constant transition probabilities) approach, [St].

Most people assume, without thinking about it, that the mathematics which was developed in the past is automatically appropriate for all new applications. Yet experience since the Middle Ages has repeatedly shown that new applications often require modifications to the classically taught mathematics. Probability is no exception; probability arose long after the formation of the classical mathematics and it is not obvious that we should try to use the unmodified mathematics in all applications. We need to examine our beliefs, and the appropriateness of the usual mathematical model, before we adopt it completely. It may well be that the usual models of probability need various modifications of mathematics.

Exercises 4.8

4.8-1   Apply the Bernoulli value to the St. Petersburg paradox.

4.8-2   Find the generating function of Example 4.8-3.

4.8-3   Do a computer simulation of Example 4.8-3.

4.9   Summary

In this Chapter we have examined the simplest infinite sample spaces which seem to arise naturally. We found that there is a uniform method which is based on the finite state diagram representation of the problem. This method leads to rational generating functions, from which you can easily get both the mean and the variance of the distribution.

If you can factor the characteristic polynomial then you can get, via partial fractions and dividing the individual terms out, the actual terms of the generating functions in terms of sums of powers of the characteristic roots. But the values of the solution are rational in the coefficients and initial conditions, while generally speaking the characteristic roots are algebraic! It follows that the algebraic part of the solution must in fact cancel out, that somehow the sums of the roots that occur must be expressible in terms of the coefficients of the characteristic equation and products of the transition probabilties, namely by simply dividing out the rational fraction.

We introduced the safety device of the *conservation of probability*; it involves one extra state (which you generally do not care about though it may be needed in some problems). It also provides an intuitive way of thinking about problems as if probability were flowing water.

We also warned you that even this simple model can give some startling results if you try to apply standard mathematics without careful thinking; the Castle Point paradox was designed to dramatize the effect. The finite state diagram provides some safety in these matters.

Appendix 4.A   *Linear Difference Equations*

Linear difference equations with constant coefficients are treated almost exactly the same way as are linear differential equations with constant coefficients (because of the common underlying linearity). Therefore we first review the theory of linear differential equations with constant coefficients.

For the *linear differential equation with constant (real) coefficients* (where $y^{(n)}$ means the nth derivative of y)

$$a_0 y^{(n)}(x) + a_1 y^{(n-1)}(x) + \cdots + a_n y(x) = f(x) \qquad (4.\text{A}-1)$$

the knowledge of any particular integral of (4.A–1), call it $Y(x)$, reduces the problem to the *homogeneous equation*

$$a_0 y^{(n)}(x) + a_1 y^{(n-1)}(x) + \cdots + a_n y(x) = 0 \qquad (4.\text{A}-2)$$

The *eigenfunctions* of this homogeneous equation are essentially the exponentials ("essentially" because of the possibility of multiple roots in the characteristic equation)

$$y(x) = e^{mx} = \exp(mx)$$

When this function is substituted into (4.A–2) we get the characteristic equation, since the exponential cancels out (as it must since we used eigenfunctions),

$$a_0 m^n + a_1 m^{n-1} + \cdots + a_n = 0 \qquad (4.\text{A}-3)$$

This equation has the n *characteristic roots*

$$m_1, \quad m_2, \quad m_3, \quad \ldots, \quad m_n$$

If all the characteristic roots are distinct then the general solution of the homogeneous equation is

$$y(x) = \sum_{k=1}^{n} C_k \, e^{m_k x} \qquad (4.\text{A}-4)$$

For a double root you try both $x \exp(mx)$ and $\exp(mx)$, for triple roots $x^2 \exp(mx)$, $x \exp(mx)$ and $\exp(mx)$, etc. If a root is complex then since we are assuming that the coefficients of the original differential equation are real the conjugate root also occurs. Let the roots be

$$a + ib \quad \text{and} \quad a - ib$$

In place of the two complex exponentials you can, if you wish, use instead

$$e^{ax} \{ C_1 \cos bx + C_2 \sin bx \} \qquad (4.\text{A}-5)$$

and multiple roots are treated just as in the case of simple roots (extra factors of x as needed). If we add the particular integral we assumed that we knew, $Y(x)$, to the general solution of the homogeneous equation then we have the complete solution of the original equation (4.A–1).

For the *linear difference equation*

$$a_0 y(n + m) + a_1 y(n + m - 1) + \cdots + a_n y(n) = f(n) \qquad (4.A–6)$$

we again assume that we have a solution (integral) $Y(n)$ of the complete equation and we are then reduced, as before, to the homogeneous equation

$$a_0 y(n + m) + a_1 y(n + m - 1) \cdots + a_n y(n) = 0 \qquad (4.A–7)$$

In place of the eigenfunctions $\exp(mx)$ we conventionally write (x becomes n of course)

$$e^{mn} = r^n \qquad \text{and} \qquad r = e^m \qquad (4.A–8)$$

and we get the characteristic equation

$$a_0 r^n + a_1 r^{n-1} + \cdots + a_n = 0 \qquad (4.A–9)$$

The general solution of the homogeneous equation is then of the form

$$y(n) = \sum_{k=1}^{n} C_k r_k^n \qquad (4.A–10)$$

and the rest is the same as for differential equations with the slight difference that multiple roots lead to

$$r^n, \qquad nr^n, \qquad n^2 r^n, \ldots$$

To illustrate the solution of a linear difference equation with constant coefficients we choose the perhaps best known example, that of the Fibonacci numbers which are defined by

$$f_{n+1} = f_n + f_{n-1}, \qquad f(0) = 0, \qquad f(1) = 1$$

Each Fibonacci number is the sum of the previous two numbers.

Set $f_n = r^n$ and we are led to the characteristic equation

$$r^2 - r - 1 = 0$$

whose solutions are

$$r_1 = \frac{1 + \sqrt{5}}{2}, \qquad r_2 = \frac{1 - \sqrt{5}}{2}$$

Therefore the general solution is

$$f_n = C_1 r_1^n + C_2 r_2^n$$

Using the initial conditions we get the two equations

$$f_0 = 0 = C_1 + C_2$$
$$f_1 = 1 = C_1 r_1 + C_2 r_2$$

Solving for these unknown coefficients C_i we get, after a little algebra,

$$f_n = \frac{1}{\sqrt{5}} \left[\left(\frac{1+\sqrt{5}}{2} \right)^{n-1} - \left(\frac{1-\sqrt{5}}{2} \right)^{n-1} \right]$$

Upon examinating how $\sqrt{5}$ enters into the square brackets we see, on expanding the binomials, that only the odd powers remain and that the front radical removes all quadratic irrationalities, as it must since from the definition all the Fibonacci numbers must be integers.

For the nonhomogeneous equation the particular integral of the original equation, $Y(x)$ or $Y(n)$ in the two cases, is *usually* found by plain, simple guessing at the general form, substituting into the equation, and finally determining the unknown coefficients of the assumed form using the linear independence of the terms to justify the equating of the coefficients of like powers of t.

5

Continuous Sample Spaces

5.1 A Philosophy of the Real Number System

It is natural to ask some one to pick an equally likely random point on the unit line segment. By the symmetry that most people perceive, or because nothing else was said, we see that the probability of picking any point must be the same as that of any other point. If we do not assign this value to be $p = 0$ then the total probability would be infinite. But apparently picking the probability to be zero for each point would lead to a zero total probability. Thus we are led to examine the usual mathematical answer that the real line is composed of points, having no length, but also that the unit line segment has length 1.

Indeed, up to now we have evaded, even in the infinite discrete probability space, a number of difficulties. For example you cannot pick a random (uniform) positive integer, because whatever number you name, "almost all" positive integers (meaning more than any fraction less than 1 you care to name) will be greater than the one you picked! If we choose to pick the positive integer n with corresponding probability

$$p(n) = 6/(\pi n)^2$$

then the probability would total 1 (since $\Sigma 1/n^2 = \pi^2/6$). By and large in the previous chapter we dealt *only* with naturally ordered infinite sequences of probabilities; we did not venture into situations where there was no natural ordering. Thus we did not assign a probability distribution over all the rational numbers between 0 and 1. It can be done, but the solution is so artificial as to leave doubt in one's mind as to the wisdom of its use in practice.

We have again raised the question of the relevance for probability theory of the usual real number system that was developed long before probability theory was seriously considered as a field of research. The real number system is usually developed from the integers ("God made the integers, the rest is man's handiwork." Kronecker), first by the extension to the rational numbers, and then by the extension to all real numbers via the assumption that *any* (Cauchy) convergent sequence of numbers must have a number as the limit. In the late middle ages the progress of physics found that the real numbers did not serve to describe forces, and the concept of vectors was introduced. Later the concept of a tensor was needed to cope with naturally occuring physical concepts. Thus without further examination it is not completely evident that the classical real number system will prove to be appropriate to the needs of probability. Perhaps the real number system is: (1) not rich enough—see non-standard analysis; (2) just what we want—see standard mathematics; or (3) more than is needed—see constructive mathematics, and computable numbers (below).

In the theory of computation, the concept of a *computable number* has proved to be very useful. Loosely speaking, a computable number is one for which a definite algorithm (process) can be given for computing it, digit by digit. The computable numbers include all the familiar numbers, including e, π, and γ. If by "describe" is meant that you can say how one can compute it digit by digit, then the computable numbers include all the numbers you can ever hope to describe. The set of computable numbers is countable because every computable number can be computed by a program and the set of programs is countable (programs are merely finite sequences of 0s and 1s).

What are all these uncountably many noncomputable numbers that the conventional real number system includes? They are the numbers which arise because it is assumed that "the limit of any Cauchy convergent sequence has a limiting number"; if you say, "the limit of any convergent sequence of describable numbers has a limit" then none of these noncomputable (uncountably many) numbers can arise. As an example of the consequences of what is usually assumed, take all the real numbers between 0 and 1 in the standard real number system, and remove the computable numbers. What is left is a noncountable set, no member of which you can either name or give an algorithm to compute it! You can never be asked to produce one of them since the particular number asked for cannot be described! Indeed, if you remove the computable numbers from each (half-open) interval (set) $[k, k + 1)$, then according to the usual axiom of choice you can form a new set by selecting one number from each of these sets (whose members you can never describe!) and form a new set. How practical is this set? Is this kind of mathematics to be the basis of important decisions that affect our lives?

The intuitionists, of whom you seldom hear about in the process of getting a classical mathematical education, have long been articulate about the troubles that arise in the standard mathematics, including the paradoxes, in the usual foundations of mathematics. One such is the Skolem-Lohenheim paradox which asserts that any finite number of postulates has a realization that is countable. This means that no finite number of postulates can uniquely characterize the accepted real number system. When we apply the Skolem-Lohenheim paradox to the real number system we observe that (usually) the number of postulates of the real number system is finite hence (apparently) any deduction from these alone must not be able to prove that the set is noncountable (since any deduction from them alone must apply to *all* realizations of the model)! Again, the Banach-Tarski paradox, that a sphere can be decomposed into a finite number of congruent parts and reassembled to be a complete sphere of any other radius you choose, a pea size or the size of the known universe, suggests that we must be wary of using such kinds of mathematics in many real world applications including probability theory. These statements warn us that we should not use the classical real number system without carefully thinking whether or not it is appropriate for new applications to probability.

We should also note that isolated finite values of functions are traditionally ignored; thus

$$\int f(x)\,dx \qquad \text{with} \qquad |f(x)| \leq M$$

over the intervals $(a < x < b)$ and $(a \leq x \leq b)$ have the same value. The isolated end values do not affect the result; indeed any finite number of (finite) missing values, or discontinuities, cannot affect the result (where we, of course, are assuming reasonable functions). The Riemann integral generally handles all the reasonable functions one can expect to meet when dealing with reality.

What are we to think of this situation? What is the role *in probability theory* for these numbers which can never occur in practice? Do we want to take over the usual real number system, with all its known confusions, when we have enough of our own? On the other hand, the real number system has the advantage that it is well known, (even its troubles are familiar), fairly well explored, and has comparatively few paradoxes left. As we go on and find further paradoxes in our probability work we need to ask whether the trouble arises from the nature of the assumed real number system or from the peculiar nature of the probability theory. See Section 8.5 for much the same argument.

5.2 Some First Examples

Rather than begin with some assumptions (postulates if you prefer) whose source, purposes, and reasonableness are not clear, we will begin by doing a couple of problems in a naive way so that you will see why probability theory has to make certain assumptions.

Example 5.2–1   *Distance From a River*

Suppose you are on the bank of a straight river and walk a kilometer in a randomly chosen direction. How far from the river will you be? Figure 5.2–1.

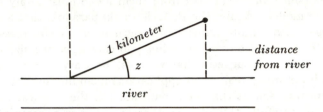

Distance from river

FIGURE 5.2–1

By implication we are restricted to directions away from the river, and hence angles z between 0 and π. By symmetry we see that we can restrict ourselves to angles between 0 and $\pi/2$, see Figure 5.2–1. It is natural to assume that all angles in the interval are equally likely to be chosen (because nothing else was said). If the angle is chosen to be z then we next have to think about the integral (rather than the sum) of all angles, and since the total probability must be 1 we have to *normalize* things so that the probability of choosing an angle in an interval Δz is

$$(2/\pi)\Delta z.$$

hence the total probability will be

$$\int_0^{\pi/2} (2/\pi)\, dz = 1$$

as it should.

Now to get the expected distance from the river we note that for the angle z the final distance is $\sin z$ (where z is the angle between the direction and the river). Hence the expected distance is the expectation of the variable $\sin z$,

$$E\{\sin z\} = \int_0^{\pi/2} (2/\pi)\sin z\, dz$$

$$= (2/\pi)(-\cos z)\ \Big|_0^{\pi/2} = 2/\pi = 0.6366\ldots$$

Example 5.2–2   Second Random Choice of Angle

Suppose as in the previous example you walk the kilometer. At that point you choose a new random direction and then you begin to walk another kilometer. What is the probability you will reach the river bank before the second kilometer is completed?

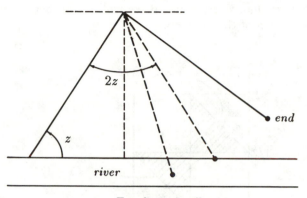

Two legs of walk

FIGURE 5.2–2

Figure 5.2–2 shows the situation, and you see that you will reach the bank only when the new angle lies within an arc that is twice the size of the original angle. We now have to integrate over all angles 0 to 2π for the second trial, and the successes would be over an arc of size $2z$. The *normalization factor* is now $1/2\pi$ and the integration of the new angle gives $2z$, so that we have success for angle z of probability z/π. This is now to be integrated over all z. In mathematical symbols

$$\int_0^{\pi/2} (2/\pi)(z/\pi)\, dz = (2/\pi^2)(z^2)/2 \Big|_0^{\pi/2}$$

$$= (2/\pi^2)(\pi/2)^2/2 = 1/4$$

which seems reasonable. See Exercise 5.2–9 for the sample space.

Example 5.2–3   *Broken Stick*

We break a stick at random in two places, what is the probability that the three pieces will form a triangle?

Naturally we assume that each break occurs uniformly in the possible range. Since it is only the relative lengths that matter we can assume that the original stick is of length 1. Let the two breaks be at points x and y, each uniformly in the unit interval $(0,1)$. We could use symmetry and insist that $x \leq y$, but rather than worry about it we will not bother to think carefully about this reduction of the problem.

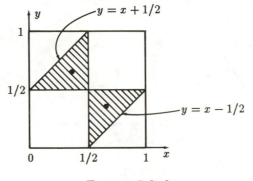

FIGURE 5.2–3

The sample space of the problem is the unit square, Figure 5.2–3. If the breaks, both x and y, are at places greater than $1/2$ then the other side of the triangle would be more than $1/2$ and no actual triangle is possible, (no side of a triangle can be greater than the sum of the other two sides), and similarly if both breaks are at places less than $1/2$. Thus the upper right and lower left are regions where the pieces could not form a triangle. The lines $x = 1/2$ and $y = 1/2$ are thus edges of regions that need to be investigated. We also require that the difference between x and y not exceed $1/2$. The equations

$$y - x = 1/2 \quad \Longrightarrow \quad y = x + 1/2$$
$$x - y = 1/2 \quad \Longrightarrow \quad y = x - 1/2$$

are the lines that are the edges of similar impossible regions, as shown in the Figure. To decide which of these smaller regions will let us form triangles from the three pieces, we observe that if $x = 1/3$, $y = 2/3$, or if $x = 2/3$, $y = 1/3$ then from the pieces we can make an equilateral triangle, hence these points fall in the feasible regions. The area of each is $1/8$, so that the total probability of successes is (we started with unit area)

$$P = 1/4$$

Is this answer reasonable? Let x fall in either half of the length; half the time y will fall there too, and no triangle will be possible, so already we see that the probability must be less than $1/2$. If the two places of the breaks, x and y, are in opposite halves of the stick, still they must not be too far apart. Hence the $1/4$ seems reasonable.

Example 5.2–4   The Buffon Needle

The famous French naturalist Buffon (1707–1788) suggested the following Monte Carlo determination of the number π. Imagine a flat plane ruled with parallel lines having unit spacing. Drop at random a needle of length $L \leq 1$ on the plane. What is the probability of observing a crossing?

"At random" needs to be thought about. It *seems* to mean that the center of the needle is uniformly placed with respect to the set of lines, and that the angle θ that the needle makes with the direction of the lines is uniform, and by symmetry we might as well suppose that the angle is

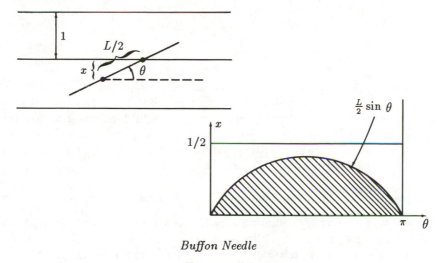

Buffon Needle

FIGURE 5.2–4

uniform in the interval $(0, \pi)$. Figure 5.2–4 shows the situation. There will be a crossing if and only if the distance x from the center of the needle to the nearest line is less than $L/2 \sin \theta$. Thus the boundary in the sample space between success and failure is the curve

$$x = (L/2) \sin \theta$$

Since the angle runs between 0 and π the area for success is

$$A = \int_0^\pi (L/2) \sin \theta \, d\theta = (L/2)(-\cos \theta) \Big|_0^\pi$$

$$= L$$

The total area of sample space of trials is $\pi/2$, so the ratio of the success area to the total sample space area is

$$2L/\pi$$

We can make some trials of tossing a needle and get an estimate of the definite number π. Since it is wise to have the chance of success and failure comparable, we picked $L = 8/10$ and ran the following trials. We see that the first two 100 trials in Table 5.2–2 are unlikely, but gradually the results are swamped by the law of large numbers (see Section 2.8).

TABLE 5.2–1

Successes	Total	Estimate of π
43	100	3.7209
87	200	3.6782
142	300	3.3803
192	400	3.3333
246	500	3.2520

Do not think that running many cases will produce an arbitrarily good estimate of π; it is more a test of the particular random number generator in the computer! Thus these Monte Carlo runs are useful for getting reasonable estimates, and for checking a computation to see that gross mistakes have not been made, but they are not, without careful modification, an efficient way of computing numerical constants like π.

Example 5.2–5   Another Monte Carlo Estimate of π

There is a simpler way of estimating π by Monte Carlo methods than Buffon's needle. Suppose we take two random numbers x_i and y_i between 0 and 1 and regard them as the coordinates of a point in the plane. The probability that the random point will lie within the arc of the circle, $(x \geq 0, y \geq 0)$

$$x^2 + y^2 = 1$$

is exactly $\pi/4$ (see Figure 5.2–5).

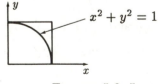

FIGURE 5.2–5

To check the model we simulated the experiment on a small computer. The variance of the Binomial trials is $\{npq\}^{1/2}$, and this leads to

$$\sqrt{\pi p q} = \sqrt{\frac{\pi}{4}\left(1 - \frac{\pi}{4}\right)}\, n = 0.411\ldots\sqrt{n}$$

The following table is the results.

TABLE 5.2–2

Trials	Experiment			Theory	
	Successes	π(Estimate)		Successes	$\sqrt{npq}$
100	82	3.28		78.5	4.11
200	160	3.20		157.1	5.81
300	245	3.2667		235.6	7.11
400	323	3.23		314.2	8.21
500	407	3.256		392.7	9.18
600	482	3.2133		471.2	10.1
700	557	3.1829		549.8	10.9
800	636	3.18		628.3	11.6
900	707	3.1422		706.9	12.3
1000	788	3.152		785.4	13.0

The table indicates that the experimental data is close to the theoretical values and is generally within the likely distance from the expected value given in the last column. Thus we have again verified the model of probability we used.

Notice that in the Examples 5.2–1 and 5.2–2 we normalized each random variable while in Example 5.2–3 we merely took the ratio of area of the successes to the total area. It is necessary that you see that these two methods are equivalent. The normalization at each stage adds, perhaps, some clarity and security since you have to think oftener, but taking the ratio is often the simplest way to get the probability of a success—always assuming that the probability distribution is uniform. If it is not then the simple modification of including the weight factor in the integrals is obvious and need not be discussed at great length.

Example 5.2–6   *Mean and Variance of the Unit interval*

Given the continuous probability distribution

$$p(x) = 1 \qquad (0 \le x \le 1)$$

what is the mean and variance?

The mean is given by (sums clearly go into integrals)

$$\mu = \int_0^1 x\, p(x)\, dx = \int_0^1 x\, dx = 1/2$$

For the variance we have, similarly,

$$\sigma^2 = \int_0^1 (x - 1/2)^2 p(x)\, dx$$

$$\int_0^1 (x - 1/2)^2\, dx = (x - 1/2)^3/3 \Big|_0^1 = 1/12$$

Exercises 5.2

5.2–1   A board is ruled into unit squares and a disc of a radius $a < 1$ is tossed at random. What is the probability of the disc not crossing any line? Plot the curve as a function of a. When is the chance 50–50? Ans. $p = (1 - a)^2$, $a = 1 - 1/\sqrt{2}$.

5.2–2   Same as 5.2–1 except the board is ruled into equilateral triangles of unit side.

5.2–3 Same as 5.2–1 except regular hexagons.

5.2–4   Given a circle A of radius a with a concentric circle B of radius $b < a$, what is the probability that a coin of radius $c (c < b)$ when tossed completely in A will not touch the edge of circle B? What is the 50–50 diameter? Ans. If $a > b + 2c$ then $P = \{(a + c)^2 - 4bc\}/a^2$, if $a \le b + 2c$ then $P = (a - c)^2/a^2$.

5.2–5   Analyse the Buffon needle problem if you assume that one end of the needle is uniformly at random instead of the middle.

5.2–6   A circle is inscribed in a square. What is the probability of a random point falling in the circle?

5.2–7   A sphere is inscribed in a cube. What is the probability of a random point falling in the sphere?

5.2–8   In Example 5.2–6 the square is inscribed in another circle what is the probability of being in the square and not in the inner circle?

5.2–9   Sketch the sample space of Example 5.2–2.

5.2–10   Two people agree to meet between 8 and 9 p.m., and to wait $a < 1$ fraction of an hour. If they arrive at random what is the probability of meeting? Ans. $a(2 - a)$.

5.2–11   A piece of string is cut at random. Show that the probability that ratio of the shorter to the longer piece is less than $1/3$ is $1/2$.

5.2–12   A random number is chosen in the unit interval. What is the probability that the area of a circle of this radius is less than 1? Ans. $1/\sqrt{p}$.

5.2–13 The coefficients are selected independently and uniformly in the unit interval. What is the probability that the roots of

$$ax^2 + 2bx + c = 0$$

are real? Ans. $5/9$.

5.2–14   Same as 5.2–13 except the equation $ax^2 + bx + c = 0$. Ans. $(5 + 6ln2)/36 = 0.2544\ldots$

5.2–15   As in Example 5.2–1 you pick a random angle, but this time you pick a distance selected uniformly in the range $0 < x < 1$. What is the expected distance from the river?

5.2–16   Two random numbers are selected from the interval $0 < x < 1$. What is the probability that the sum will be less than 1?

5.2–17   Three random numbers are selected uniformly from the interval $0 < x < 1$. What is the probability that the sum will be less than 1? Ans. $1/3!$.

5.2–18   Select two random numbers from the interval $0 < x < 1$. What is the probability that the product is less than $1/2$?

5.2–19   If a point is selected in a circle with a probability density proportional to the radius, what is the normalizing factor?

5.2–20   Same as Exercise 5.2–19 except for a sphere.

5.2–21   Same as Exercise 5.2–17 except n numbers are selected. Ans. $1/n!$.

5.3 Some Paradoxes

We have been fairly glib about the assignment of a uniform probability in a finite interval. As the following examples show, the troubles that can arise are more subtle than most people initially think. See also [St].

Example 5.3–1   *Bertrand's Paradox*

The probabilist Bertrand asked the question, "If you draw a random chord in a circle what is the probability that it is longer than the side of the equilateral triangle that you can draw in the same circle?" The equilateral triangle scales the problem so that the probability is independent of the radius of the circle.

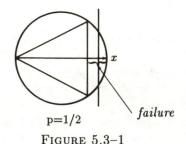

p=1/2

FIGURE 5.3–1

There are three different ways of reasoning. In the first, Figure 5.3–1, we select the position where the chord cuts the radius that is perpendicular to the chord, and *assume* that this is the equally likely choice. We see, from the geometry, that the probability is exactly 1/2.

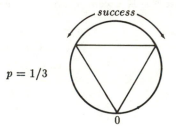

$p = 1/3$

FIGURE 5.3–2

The second way is to suppose that one point on the circumference is picked arbitrarily, and *assume* that the other intersection is uniformly spread around the circumference. From Figure 5.3–2 we see that the probability is exactly 1/3.

The third way we *assume* that the middle of the chord is spread uniformly around the inside of the circle. From Figure 5.3–3 we see that the probability is 1/4.

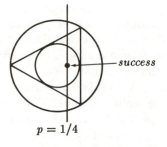

$$p = 1/4$$

FIGURE 5.3–3

This is a famous paradox, and shows that when the chord is chosen at random the question of what is to be taken as uniform is not as simple as one could wish. The three different answers, 1/2, 1/3 and 1/4, can all be defended, and they are not even close to each other!

Suppose we make the *apparently* reasonable proposal that if A is random with respect to B, then B is random with respect to A—and we hope that the "uniform assignments" also agree. We therefore invert the Bertrand random chord tossing, and think of a line as fixed and toss a circle. If an intersection is to occur then the center of the circle should occur uniformly in the area about the line. To be concrete we imagine a line segment of finite length, and toss a coin of unit radius on the floor. We count only those trials where the coin crosses the line *and* forms a chord. We see that the chord will be longer than the corresponding triangle when the center is within a distance of 1/2 of the radius to the line, and have in Figure 5.3–4 allowed for the end effects of the finite line. As the line gets longer and longer, the end effects decrease

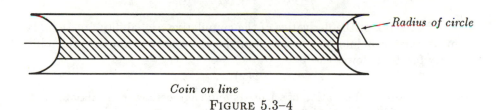

Coin on line

FIGURE 5.3–4

relative to the rest of the experiment, so we look at the main part to see that the probability will approach 1/2. Thus we get one of Bertrand's answers. You may think that this is the "correct answer," but it shows that even the principle "A random with respect to B implies B random with respect to A" does not carry over to the words "uniformly random." Indeed, if A is always 1 and B is a random number chosen uniformly in ($0 < x < 1$), then B is random with respect to A, but A is not random with respect to B.

The Bertrand paradox is not presented merely to amused you; it is something for you to worry about if you ever plan to model the real world by a probability model and then act on the results. Just how will *you* choose the assignment of the probabilities, since different assignments can give significantly different results? In computer simulations all too often the actual randomness is left to some programmer who is interested in programming and not in the relevance of the model to reality. Because of this, most simulations that use random numbers are suspect *unless* a person with an intimate understanding of the real situation being modeled examines the way the randomness enters into the computation. There is, so far as the author knows, no simple solution to the situation that the assignment of the random choice may not be left to the programmer to do the obvious; every assignment of a probability distribution must be carefully considered if you are to act responsibly on the results of a simulation. Symmetry was used earlier to justify the uniform distribution of a random variable; later in Chapters 6, 7 and 8 we will give further discussions of how to choose the initial random distributions.

Apparently we cannot talk with any mathematical rigor about the probability of choosing a random point, but we can *accumulate* the probability in intervals. In particular we define *the cumulative probability* between $-\infty$ and x as

$$\Pr\{-\infty \leq t \leq x\} = P(x) = \int_{-\infty}^{x} p(t)\, dt$$

for some probability density distribution p(x) where, of course,

$$\int_{-\infty}^{\infty} p(t)\, dt = 1$$

Thus the probabilty of falling in an interval (a, b) is

$$\int_{a}^{b} p(t)\, dt = P(b) - P(a)$$

If we think of the derivative of the cumulative probability function (except at isolated points for practical functions) we have again the probability density function (lower case)

$$p(x)$$

In the infinitessimal notation of scientists and engineers we can talk about the probability of being in an interval dx as being

$$p(x)\, dx$$

but not about the probability at a point. Talking about the probability density $p(x)$ is very convenient, but careful thinking requires us to go to the integral form whenever there is any doubt about what is going on.

We now turn to a classic problem due to Lewis Carroll (Charles Dodson, the Oxford mathematician) which illustrates the danger of using the idea of a random point in a plane.

Example 5.3–2   *Obtuse Random Triangles*

Lewis Carroll posed the problem: if you select three random points in a plane what is the probability that the triangle will be obtuse?

For a long time the standard reasoning was to place the x-axis of a cartesian coordinate system along the longest side of the triangle, and the origin at the vertex joining the second longest side. We will scale the longest

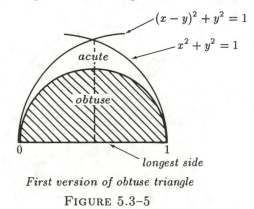

First version of obtuse triangle

FIGURE 5.3–5

side to be of unit length since it is only the ratio of areas that will matter. Then looking at the Figure 5.3–5 we see that the third point must fall inside the circle of unit radius about the origin. But the point must also fall within a similar circle centered about the other end of the longest side. We now remember that the angle in a semicircle is a right angle, hence the circle of half the radius marks the boundary between obtuse and acute angles. The figure shows the region of obtuse angles. By symmetry it is enough to use only the upper half of the plane. The area of all the possible triangles is given by

$$\text{Area} = 2 \int_{1/2}^{1} \sqrt{(1 - x^2)} \, dx$$

Using the standard substitution of $x = \sin t$ we get

$$\text{Area} = 2 \int_{\pi/6}^{\pi/2} \cos^2 t \, dt$$

Using the half angle substitution we have

$$\text{Area} = \int_{\pi/6}^{\pi/2} (1 + \cos 2t) \, dt$$

$$= \{\pi/2 - \pi/6\} + (1/2)\{\sin \pi - \sin \pi/3\}$$

$$= \pi/3 - \frac{\sqrt{3}}{4}$$

Now the area where the obtuse triangles occur is clearly $\pi/8$, so we have the probability of an obtuse triangle is

$$\frac{\frac{\pi}{8}}{\frac{\pi}{3} - \frac{\sqrt{3}}{4}} = 0.639\ldots$$

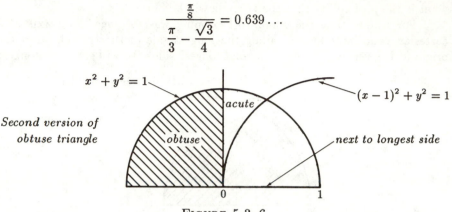

$x^2 + y^2 = 1$

acute

$(x-1)^2 + y^2 = 1$

Second version of obtuse triangle

obtuse

next to longest side

0   1

FIGURE 5.3–6

This solution was widely accepted until someone thought to compute the answer taking the x-axis along the second longest side with unit length, and the vertex at the junction of that side with the shortest side. From Figure 5.3–6 we see that the longest side forces the other vertex to be outside a circle of radius 1 about the point $x = 1$, but that it must also lie inside a circle of unit radius about the origin since the shortest side is less than 1. The Figure shows the obtuse triangle area as shaded, and it is only the contiguous small area to the right of the y-axis that has acute angles. For this small area we have the integral

$$\int_0^{1/2} \left[\sqrt{(1 - x^2)} - \sqrt{\{(1 - (1-x)^2\}} \right] \, dx$$

In the first integral we use again $x = \sin t$, and in the second integral we use $1 - x = \sin t$. We get

$$\int_0^{\pi/6} \cos^2 t \, dt - \int_{\pi/6}^{\pi/2} \cos^2 t \, dt$$

Again using the half angle substitution we get

$$\frac{\sqrt{3}}{4} - \frac{\pi}{12}$$

for this small sector. Hence the probability of an obtuse angled triangle is

$$(\text{Area of success})/(\text{total area}) = \frac{\frac{\pi}{4}}{\frac{\pi}{6} - \frac{\sqrt{3}}{2}} = 0.821\ldots$$

This is very different from the first answer, and a simple look at the two figures will show that the two probabilities are indeed quite different.

What is wrong? It is not the choice of coordinate systems, it is the idea that you can choose a random point in the plane. Where is such a point? It is farther out than any number you can mention! Hence the distance between the points is arbitrarily large—the triangle is practically infinite and the assumptions of the various steps, including supposing that the side chosen is of unit length, are dubious to say the least.

It is significant that the first standard solution offered for the Lewis Carroll problem was accepted for so long, and it indicates the difficulty of using probability in practice, especially continuous probability. You need to be very careful whenever you come in contact with the infinite.

If we attempt to do a Monte Carlo simulation by choosing points in a square of sides $2N$, for example, and letting the length of the side approach infinity we find from the random number generator we get a number x between 0 and 1, and to get to the interval $-N$ to N we must use

$$(2x - 1)N$$

By rearranging things a bit we see that we will always (for the uniform distribution) reduce the problem to the unit square and the factor N will drop out. For this situation we would get an answer for the probability of an obtuse triangle. But if we had chosen a circle instead of the square we would have gotten a different answer! If the difference between the answers for the square and circle is not obvious, then consider a long, flat rectangle and let the size approach infinity while keeping the "shape" of the rectangle constant. The probability of an obtuse triangle in a long, low rectangle is very high and depends on the shape of the rectangle. Hence, just which shape should we choose to let approach infinity? The answer clearly depends on this choice of shape and not on the size of the figure.

Example 5.3–3   *A Random Game*

In this game I choose a random number x uniformly between 0 and 1. You then begin to pick uniformly distributed random numbers y_i from the same unit interval and are paid one unit of money for each trial until your y_i is at least as large as my x, Figure 5.3–7. If it is to be a fair game then what price should I ask? Alternately, how many trials can you expect to make until your $y_i \geq x$?

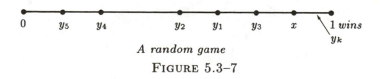

A random game

FIGURE 5.3–7

For any given x that I pick, what is the probability that you will get a number y_i larger that x and hence end the game? Since we are assuming a uniform distribution, the probability is just the length of the interval for failure. It is simply

$$\text{Prob}\{y_i \geq x\} = 1 - x$$

The expected number of trials for any fixed x (other than $x = 1$) is the reciprocal of the probability of the event (see 4.2–3); hence your expected gain is, for my y,

$$1/(1 - x) - 1 = x/(1 - x)$$

To get the expected value *for the game* we average this over all possible choices of my x

$$\int_0^1 \frac{x}{1 - x}\, dx = -x - ln(1 - x) \Big|_0^1 = \infty$$

Few people would play this fair (?) game with the infinite entry fee!

We can examine how this happens by dropping back to finite problems. There are several ways of doing this, and we shall examine two of them.

Suppose first that there are only a finite number of numbers in the unit interval, say N equally spaced numbers ($0 \leq x < 1$). The numbers are therefore $0/N, 1/N, 2/N, \ldots, (N-1)/N$. I must allow your sequence of numbers selected from m/M, ($m = 0, 1, 2, \ldots N - 1$) to end on an equality since otherwise if my x is the largest number, $(N-1)/N$, then you could never exceed it and you would be paid forever! For my given $x = k/N$, ($k = 0, 1, \ldots, N - 1$), you have the probability

$$(N - k)/N$$

of getting a single choice at least as large as my x. The expected number of trials will be for my particular choice x

Expected number of trials before a failure $= N/(N - k) - 1$

Summed over all my x, meaning k, we get

Average expected number of trials

$$= \frac{1}{N} \sum_{k=0}^{N-1} \left\{ \frac{N}{N - k} - 1 \right\}$$

If we set $k' = N - k$ in the first sum, then

Average expected number of trials $= \sum_{k'=1}^{N} \frac{1}{k'} - 1$

We can, by the result in Appendix 1.A, approximate the sum by the midpoint integration formula to get

$$(1/N) \int_{1/2}^{N+1/2} \frac{1}{x} \, dx - 1 \sim ln(2N+1) - 1$$

As the number of points, N, used in the unit interval increases, the expected number of trials goes to infinity like $ln\,N$. This is much like the St. Petersburg paradox (Section 4.8). We also note that as N approaches infinity we are measuring the interval with a spacing $1/N$ and we do not involve the discrete vs. the continuous.

A second approach is to choose my number from some fraction of the unit line, say $(0 \le x \le a < 1)$, which amounts to taking my number from $(0 \le x \le 1)$ and multiplying by $a < 1$. Now your y_i must exceed my number **ax**. We are led to the integral

$$\frac{1}{a} \int_0^a \frac{1}{1-x} \, dx = \frac{1}{a} \, ln \left\{ \frac{1}{1-a} \right\}$$

and as $a \to 1$ this again heads for infinity.

Again we see that even in simple problems you get results that you would not act on in the real world. These particular paradoxical results are easy to spot because they involved an infinity that is in the open; how can you hope to spot similar errors when the results are not so clearly displayed?

Probability applications are dangerous, but in practice there are many situations where there is no other basis for action and where also no action is very dangerous. It is for such reasons that we continually examine the foundations of our assumptions and the way we compute. Dropping back to one or more finite models (as does the calculus) often seems to shed a great deal of light on the situation, and is highly recommended in practice where the consequences of the calculation are important. It is the actual application that indicates the finite model to use and determines your untimate actions.

5.4   The Normal Distribution

The integral

$$I = \int_{-\infty}^{\infty} e^{-x^2/2} \, dx \tag{5.4-1}$$

will occur frequently and we need to know its value even though we cannot do the indefinite integration in closed form. There is a standard trick for the evaluation. We compute the square of the integral and then convert to the double integral form

$$I^2 = \int_{-\infty}^{\infty} e^{-x^2/2} \, dx \int_{-\infty}^{\infty} e^{-y^2/2} \, dy$$

$$= \int_{-\infty}^{\infty} \int_{-\infty}^{\infty} e^{-(x^2+y^2)/2} \, dy \, dx$$

We next formally convert to polar coordinates (since the sum of the squares strongly suggests this)

$$I^2 = \int_{0}^{2\pi} \int_{0}^{\infty} e^{-r^2/2} r \, dr \, d\theta$$

$$= 2\pi \int_{0}^{\infty} e^{-r^2/2} r \, dr$$

$$= 2\pi \left(-e^{-r^2/2} \right) \Big|_{0}^{\infty} = 2\pi$$

Hence we have for (5.4–1)

$$I = \int_{-\infty}^{\infty} e^{-x^2/2} \, dx = \sqrt{2\pi} \tag{5.4-2}$$

Since this is an important result we need to examine carefully the change of the integration from repeated to double, and the change to polar coordinates. We therefore begin with the finite integral corresponding to (5.4–1)

$$I(N) = \int_{-N}^{N} e^{-x^2/2} \, dx$$

and repeat the steps. When we get to the change to polar coordinates we look at Figure 5.4–1. We get the bounds that the integral over the inner circle is less than the integral over the square, which in turn is less than the integral over the outer circle. Doing the angle integration we have

$$2\pi \int_{0}^{N} e^{-r^2/2} r \, dr < I^2(N) < 2\pi \int_{0}^{\sqrt{2N}} e^{-r^2/2} r \, dr$$

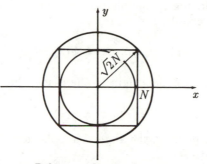

Polar coordinate change

FIGURE 5.4–1

The two end integrals can be done in closed form

$$2\pi\{1 - e^{-N^2/2}\} < I^2(N) < 2\pi\{1 - e^{-N^2}\}$$

In the limit as $N \to \infty$ each end of the inequality approachs 2π, so the middle term must too, and we are justified, this time, in the formal approach.

Example 5.4–1 *Herschel's Derivation of the Normal Distribution*

The astronomer Sir John F. W. Herschel (1792–1871) gave the following derivation of the the normal distribution. Consider dropping a dart at a target (the origin of the rectangular coordinate system) on a horizontal plane. We suppose: (1) the errors do not depend on the coordinate system orientation, (2) errors in perpendicular directions are independent, and (3) large errors are less probable than are small errors. These seem to be reasonable assumptions.

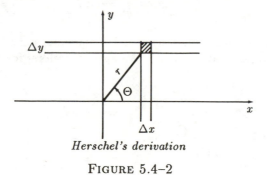

Herschel's derivation

FIGURE 5.4–2

Let the probability of the dart falling in the strip from x to $x + \Delta x$ be $p(x)\Delta x$, and of falling in the strip from y to $y+\Delta y$ be $p(y)\Delta y$, see Figure 5.4–2.

The probability of falling in the shaded rectangle is, since we assumed the independence of the coordinate errors,

$$p(x)p(y)\Delta x \Delta y = g(r)\Delta x \Delta y$$

where r is the distance of the origin to the place where the dart falls. From this we conclude that

$$p(x)p(y) = g(r)$$

Now $g(r)$ does not depend on the angle θ in polar coordinates, hence we have (using partial derivatives)

$$\frac{\partial g(r)}{\partial \theta} = 0 = p(x)\frac{\partial p(y)}{\partial \theta} + p(y)\frac{\partial p(x)}{\partial \theta}$$

From the polar coordinate relationships

$$x = r\cos\theta$$

$$y = r\sin\theta$$

we have

$$\frac{\partial p(x)}{\partial \theta} = \frac{\partial p(x)}{\partial x}\frac{\partial x}{\partial \theta} = p'(x)(-y)$$

$$\frac{\partial p(y)}{\partial \theta} = \frac{\partial p(y)}{\partial y}\frac{\partial y}{\partial \theta} = p'(y)(x)$$

This gives

$$p(x)p'(y)(x) - p(y)p'(x)(y) = 0$$

Separating variables we have the two differential equations

$$\frac{p'(x)}{xp(x)} = \frac{p'(y)}{yp(y)} = K$$

since both sides of the equation must be equal to some constant K (the variables x and y are independent). We have, therefore, for the variable x,

$$\frac{p'(x)}{p(x)} = Kx$$

$$\ln p(x) = Kx^2/2 + C$$

$$p(x) = A\,e^{Kx^2/2}$$

But we assumed that small errors were more probable than large ones, hence $K < 0$, say

$$K = -1/\sigma^2$$

and we have the normal distribution

$$p(x) = A\,e^{-x^2/2\sigma^2}$$

From the fact that the total probability must be 1 we get for the factor A

$$A = 1/(\sigma\sqrt{2\pi})$$

Similar arguments apply to the y variable.

The derivation seems to be quite plausible, and the assumptions reasonable. We will later find (Chapter 9) other derivations of the normal distribution.

Example 5.4–2 *Distance to a Random Point*

Let us select independently both the x and y coordinates of a point in the x–y plane from the same normal distribution which is centered at the origin (which is no real constraint)

$$p(t) = \frac{1}{\sigma\sqrt{2\pi}}\,e^{-t^2/2\sigma^2}$$

The expected distance from the origin is then given by

$$D = \frac{1}{\sigma^2 2\pi}\int_{-\infty}^{\infty}\int_{-\infty}^{\infty}\sqrt{(x^2+y^2)}\,e^{-(x^2+y^2)/2\sigma^2}\,dy\,dx$$

This immediately suggests polar coordinates (where we can now do the θ integration)

$$D = \frac{1}{\sigma^2 2\pi}\int_{0}^{2\pi}\int_{0}^{\infty}r^2 e^{-r^2/2\sigma^2}\,dr\,d\theta$$

$$= \frac{1}{\sigma^2}\int_{0}^{\infty}r^2 e^{-r^2/2\sigma^2}\,dr$$

Set $r = \sigma t$ to get

$$D = \sigma\int_{0}^{\infty}t^2 e^{-t^2/2}\,dt$$

Integration by parts yields easily, and use (5.4–2)

$$D = \sigma\{-te^{-t^2/2}\,\Big|_0^{\infty} + \int_{0}^{\infty}e^{-t^2/2}\,dt\}$$

$$= \sigma\sqrt{2\pi}$$

as the expected distance of the random point in the x–y plane from the origin.

This transformation from rectangular to polar coordinates needs some justification. We therefore, as before, adopt the calculus approach and start with the region $-N$ to N. From Figure 5.4–2 we see that we can easily get the bounds

$$\int_0^{2\pi} \int_0^N r^2 e^{-r^2/2\sigma^2} \, dr \, d\theta$$

$$< \int_{-N}^N \int_{-N}^N \sqrt{(x^2 + y^2)} e^{-(x^2+y^2)/2\sigma^2} \, dy \, dx$$

$$< \int_0^{2\pi} \int_0^{\sqrt{2N}} r^2 e^{-r^2/2\sigma^2} \, dr \, d\theta$$

Now as $N \to \infty$ both end integrals approach the same limit, hence the transformation from rectangular to polar coordinates is legitimate this time.

Notice that the normal distribution has the important property that the product formula $p(x)p(y)$ has circular symmetry.

Example 5.4–3   *Two Random Samples from a Normal Distribution*

Suppose we choose two random points on a line, not from a uniform infinite distribution but rather from the probability density distribution

$$p(x) = \frac{1}{\sqrt{2\pi}} e^{-x^2/2}$$

What is the expected distance between the two points?

The probability density of two independently chosen coordinates x and y from this normal distribution is their product, which is

$$\frac{1}{2\pi} e^{-(x^2+y^2)/2}$$

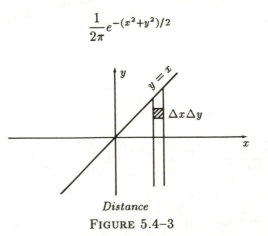

Distance

FIGURE 5.4–3

In the sample space of the events we have a probability distribution of points over the entire x–y plane to consider (although the two points lie on a line). See Figure 5.4–3. To get the expected distance between the two points we will have to consider points in the entire sample space plane, but to evaluate the integral we can take only $y > x$ and use a factor of 2 in front of the integral to compensate

$$2 \int_{-\infty}^{\infty} \int_{-\infty}^{y} \frac{1}{2\pi} (y - x) e^{-(x^2 + y^2)/2} \, dx \, dy$$

$$= \frac{1}{\pi} \int_{-\infty}^{\infty} e^{-x^2/2} \int_{x}^{\infty} y \, e^{-y^2/2} \, dy \, dx$$

$$- \frac{1}{\pi} \int_{-\infty}^{\infty} e^{-y^2/2} \int_{-\infty}^{y} x \, e^{-x^2/2} \, dx \, dy$$

where we have interchanged the limits in the first integral. Again this interchange needs to be examined, but as in the previous cases the rapid decay of the exponential term justifies the interchange. Doing the inner integrals we get (since the exponentials combine)

$$\frac{1}{\pi} \int_{-\infty}^{\infty} e^{-y^2} \, dy + \text{(same in } x\text{)}$$

$$= \frac{2}{\pi} \int_{-\infty}^{\infty} e^{-y^2} \, dy$$

But this is a known integral (5.4–2) which gives, finally,

$$2/\sqrt{\pi} = 1.128\ldots$$

as the expected distance between two points chosen from a normal distribution with unit variance.

We see from these last several examples that while the idea of a random point in the infinite plane chosen from a uniform distribution can lead to peculiar results, when we choose from a reasonably restricted distribution we get reasonable results. Example 5.3–3 is another case resembling the St. Petersburg and Castle Point paradoxes. The integral for the expected value may be infinite when you are careless - even for a finite interval!

Exercises 5.4

5.4–1   Evaluate $\int_{-\infty}^{\infty} x^n e^{-x^2/2}\,dx$.

5.4–2   In Bertrand's paradox compute the expected values of the chord in the three cases. Ans. $\pi/2 = 1.5708\ldots$, $4/\pi = 1.2732\ldots$, $4/3$

5.4–3   Why do you have trouble with the random triangle when you choose the shortest side?

5.4–4   Rework the random game if you must get a number less than mine.

5.4–5   In a circle of radius 1 the probability density function along a radius is proportional to $1/r$. What is the constant of proportionality?

5.4–6   As in Exercise 5.4–5 except proportional to r.

5.4–7   As in Exercise 5.4–5 except for a sphere.

5.4–8   As in Exercise 5.4–6 except for a sphere.

5.4–9   If $p(x) = e^{-x}$, $(0 \le x \le \infty)$, find the expected distance from the random point to the origin.

5.4–10   Same as Exercise 5.4–6 except the distance from 1.

5.4–11   Same as Exercise 5.4–7 except the point a. Ans. $a - 1 + 2\exp(-a)$.

5.4–12   What is the expected distance between two points chosen at random on a sphere? Note that you can assume that there is a great circle going through the two points, but that you need to consider elementary areas and not simply points.

5.4–13   Find the inflection points of the normal probability density function.

5.4–14 Plot the normal density function using equal sized units on both axes.

5.5   Distribution of Numbers

When we deal with numbers in science and engineering we generally use the scientific notation and then, contrary to the usual belief, the distribution of a set of naturally occuring numbers is seldom uniform, rather most of the time the probability density of seeing a number x, *where we are neglecting the exponents and looking only at the so called mantissa*, is

$$p(x) = \frac{1}{x\,ln\,10} \tag{5.5–1}$$

when we regard the numbers as lying between 1 and 10). This is naturally called *the reciprocal distribution.*

Again, we are working in the floating point (or scientific) notation that is used in almost all practical computation, and we are concerned with the decimal digits and not with the exponent. With the increasing use of computers this is an important topic having many consequences in numerical computations, including roundoff propagation studies.

Integrating the distribution (5.5–1) to get its cumulative distribution we get

$$P(x) = \frac{\ln x}{\ln 10} = \log_{10} x, \quad P(10) = 1 \tag{5.5–2}$$

From this it follows that the probability of seeing a leading digit N is

$$\Pr \text{ (leading digit is } N) = \log(N+1) - \log N = \log(1+1/N)$$

To check this experimentally we took the 50 physical constants from the NBS Handbook of Mathematical Functions (1964). We got the following table 5.5–1

TABLE 5.5–1

Leading digit	Observed	Theoretical	Difference
1	16	15	1
2	11	9	2
3	2	6	−4
4	5	5	0
5	6	4	2
6	4	3	1
7	2	3	−1
8	1	3	−2
9	3	2	1
	50	50	0

We see from this simple table that the reciprocal distribution is a reasonably accurate approximation to the physical constants tabulated in the book.

But this distribution is also a property of our number system as can be seen from the following series of results. We will want, ultimately, the distribution of a product of two numbers x and y drawn from the two distributions $f(x)$ and $g(y)$ so we examine this first.

Example 5.5–1   *The General Product*

Consider the product $z = xy$ when x is from the distribution $f(x)$ and y from $g(y)$, and where the product has been reduced to the standard floating point format. We examine the cumulative distribution $P(z)$ of the mantissa, remembering that the same number z can arise from various places in the x–y plane $(1 \leq x, y \leq 10,)$ and we need to watch especially those that come from a shift when the product is reduced to the standard floating point format. For generality we work in base b (typically $b = 2, 4, 8, 10,$ or 16), and assume that the numbers lie bewteen 1 and b (though on computers the standard notation is often from $1/b$ to 1).

When does a shift occur? Whenever the product $z \geq b$. Hence the dividing line in the sample space is clearly

$$z = xy = b$$

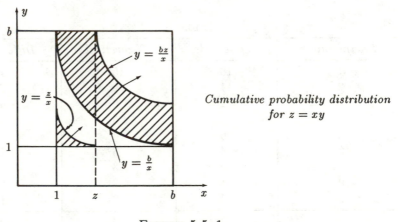

Cumulative probability distribution for $z = xy$

FIGURE 5.5–1

This is an equilateral hyperbola shown in the Figure 5.5–1. When $z < b$ we have the product $z = xy$, and when $z > b$ we have the product $z = xy/b$. There are three regions that fall below z in computing the cumulative probability function $P(z)$. These are shown in the figure. The corresponding integrals are

$$P(Z < z) = \int_1^z \int_1^{z/x} f(x)g(y)\,dy\,dx$$

$$+ \int_1^z \int_{b/x}^b f(x)g(y)\,dy\,dx$$

$$+ \int_z^b \int_{b/x}^{bz/x} f(x)g(y)\,dy\,dx$$

If $G(y)$ is the cumulative distribution of $g(y)$, that is

$$G(y) = \int_1^y g(t)\,dt, \quad G(1) = 0, \quad G(b) = 1$$

then we have, on doing the integrations with respect to y,

$$P(Z < z) = \int_1^z f(x)\{G(z/x) - G(1) + G(b) - G(b/x)\}\,dx$$

$$+ \int_z^b f(x)\{G(bz/x) - G(b/x)\}\,dx$$

$$= \int_1^z f(x)\{G(z/x) + 1 - G(b/x)\}\,dx$$

$$+ \int_z^b f(x)\{G(bz/x) - G(b/x)\}\,dx$$

To get the probability density $h(z)$ we differentiate with respect to z, remembering all the details of differentiating an integral with respect to a parameter. We get

$$h(z) = f(z)\{G(1) + 1 - G(b/z)\} - f(z)\{G(b) - G(b/z)\}$$

$$+ \int_1^z f(x)\{g(z/x)(1/x)\}\,dx$$

$$+ \int_z^b f(x)\{g(bz/x)(b/x)\}\,dx$$

$$= \int_1^z f(x)\{g(z/x)(1/x)\}\,dx \tag{5.5-3}$$

$$+ \int_z^b f(x)\{g(bz/x)(b/x)\}\,dx$$

This is the formula for the density distribution $h(z)$ for the product xy in floating point (scientific) notation. It does not look symmetric in the two variables x and y as it should since $xy = yx$, but changing variables in the integrands will convince you that the formula has the required symmetry.

Example 5.5–2   *Persistence of the Reciprocal Distribution*

If we assume that one of the variables, say y, is from the reciprocal distribution, that is

$$g(y) = 1/(y \ln b)$$

then we have for the distribution of the product (see 5.5–3)

$$
\begin{aligned}
h(z) &= \int_1^z f(x)\{x/(z \ln b)\}\{1/x\}\, dx \\
&\quad + \int_z^b f(x)\{x/(bz \ln b)\}\{b/x\}\, dx \\
&= 1/(z \ln b)\left\{\int_1^z f(x)\, dx + \int_z^b f(x)\, dx\right\} \\
&= 1/(z \ln b)
\end{aligned}
\tag{5.5–4}
$$

Thus *if one factor of the product is from the reciprocal distribution then the product has the reciprocal distribution.*

Example 5.5–3   *Probability of Shifting*

In forming a product of two numbers on a computer a shift to reduce the product to the standard format requires extra effort and possibly extra time. What is the probability of a shift? By placing the point of the number system behind or before the leading digit we can change from the probability of a shift to its complement probability.

If both numbers are from general distributions then the probability of no shift is the area under the curve $y = b/x$ (see Figure 5.5-1)

$$\Pr(\text{shift}) = \int_1^b \int_1^{b/x} f(x)g(y)\, dy\, dx$$

If both factors are from the reciprocal distribution then we have

$$\Pr(\text{shift}) = \int_1^b \int_1^{b/x} 1/(x\ln b)1/(y\ln b)\, dy\, dx$$

$$= \frac{1}{(\ln b)^2} \int_1^b (1/x)\ln y \Big|_1^{b/x}\, dx$$

$$= \frac{1}{(\ln b)^2} \int_1^b (1/x)\{\ln b - \ln x\}\, dx$$

$$= \frac{1}{(\ln b)^2} \left[\ln b \ln x - (\ln x)^2/2 \Big|_1^b\right]$$

$$= \frac{1}{(\ln b)^2} [(\ln b)^2 - (\ln b)^2/2]$$

$$= 1/2$$

Thus if both numbers come from the reciprocal distribution then it is a matter of indifference whether or not the point is before or behind the leading digit. For other distributions this is not true.

Example 5.5–4   *The General Quotient*

For the quotient

$$z = x/y$$

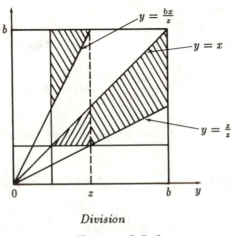

Division

FIGURE 5.5–2

we first draw the picture, Figure 5.5–2 showing the various regions and the equations bounding the regions. As before we begin with the cummulative probability distribution which is

$$H(z) = \int_1^z \int_{bx/z}^b f(x)g(y)\,dy\,dx$$

$$+ \int_1^z \int_1^x f(x)g(y)\,dy\,dx$$

$$+ \int_z^b \int_{x/z}^x f(x)g(y)\,dy\,dx$$

Again, following the earlier model we let $G(y)$ be the cummulative distribution of $g(y)$, and doing the dy integrations we get

$$H(z) = \int_1^z f(x)\{G(b) - G(bx/z) + G(x) - G(1)\}\,dx$$

$$+ \int_z^b f(x)\{G(x) - G(x/z)\}\,dx$$

Differentiating to get the probability density $h(z)$ we get

$$h(z) = f(z)\{G(b) - G(b) + G(z) - G(1) - G(z) + G(1)\}$$

$$+ \int_1^z f(x)g(bx/z)(bx/z^2)\, dx$$

$$+ \int_z^b f(x)g(x/z)(x/z^2)\, dx \qquad\qquad (5.5\text{--}5)$$

$$= \frac{b}{z^2} \int_1^z xf(x)g(bx/z)\, dx$$

$$+ \frac{1}{z^2} \int_z^b xf(x)g(x/z)\, dx$$

This is the general formula for the probability of a mantissa of the size z from a quotient $z = x/y$.

As a check on the formula suppose $g(y)$ is the reciprocal distribution

$$g(y) = 1/(y \ln b)$$

We put this into (5.4–5) to get, after a lot of cancellation,

$$h(z) = \frac{1}{z \ln b} \left\{ \int_1^z f(x)\, dx + \int_z^b f(x)\, dx \right\}$$

$$= \frac{1}{z \ln b}$$

Hence we get the reciprocal distribution so long as the denominator is from the reciprocal distribution. This increases our faith in (5.5–5).

Exercises 5.5

5.5–1   Find the formula for the probability of a shift if only one factor of a product is from the reciprocal distribution. Apply to one from the flat distribution.

5.5–2   If the numerator is 1 and the denominator comes from the reciprocal distribution find the distribution.

5.5–3   Discuss the question of a shift for quotients.

5.5–4   Compute the probability of a shift for a product if both numbers are from a flat distribution.

5.5–5   Tabulate the probability of a shift if both factors are from flat distribution for $b = 2, 4, 8, 10,$ and 16.

5.6 Convergence to the Reciprocal Distribution

How fast do we approach the reciprocal distribution from any other distribution? Is the reciprocal distribution stable in practice? These are natural questions to ask. We will first examine how, starting from the flat distribution, we approach the reciprocal distribution as we take more and more products.

Example 5.6–1   *Product of Two Numbers from a Flat Distribution*

Suppose we start with two factors from a flat distribution, that is $f(x) = 1/(b-1) = g(y)$. Then the product formula gives

$$h(z) = \frac{1}{(b-1)^2} \left\{ \int_1^z 1/x \, dx + \int_z^b b/x \, dx \right\}$$

$$= \frac{1}{(b-1)^2} \{ln \, z + b(ln \, b - ln \, z)\}$$

$$= \frac{1}{(b-1)^2} \{b \, ln \, b - (b-1) ln \, z\}$$

For base 10, for example this is

$$h(z) = \frac{10 \, ln \, 10 - 9 \, ln \, z}{81} \tag{5.6-1}$$

We can check this kind of a result by forming all possible products from the uniform distribution using 1, 2, 3, or even 4 digit arithmetic. There will be a decreasing effect of the granularity of the simulation of a uniform distribution, see Example 5.6–4.

Example 5.6–2   *Approach to Reciprocal Distribution in General*

To discuss the rapidity of the approach to the reciprocal distribution we need a measure of the distance between two probability distributions. After some thought we introduce the distance function for the distance of a probability distribution $h(z)$ from the reciprocal distribution of $r(z)$

$$\text{DIST} = D\{h(z)\} = \max_{1<z<b} \left\{ \left| \frac{h(z) - r(z)}{r(z)} \right| \right\} \tag{5.6-2}$$

This measures the maximum difference between the given distribution $h(z)$ *relative* to the reciprocal distribution $r(z)$. We showed in Example 5.5–2 that

$$r(z) = \int_1^z z f(x) \, r(z/x) \, (1/x) \, dx$$

$$+ \int_z^b f(x) \, r(bz/x) \, (b/x) \, dx$$

Subtract this from the formula (5.5–3) for $h(z)$ and divide by $r(z)$, which is *not* the variable of integration so it may be put inside the integrals

$$\frac{\{h(z) - r(z)\}}{r(z)}$$

$$= \int_1^z f(x)\{g(z/x) - r(z/x)\}/xr(z)\,dx$$

$$+ \int_z^1 f(x)\{g(bz/x) - r(bz/x)\}\{b/xr(z)\}\,dx \qquad (5.6\text{–}3)$$

But we have both

$$xr(z) = x/z\ln b = r(z/x) \quad \text{and} \quad xr(z)/b = x/bz\ln b = r(bz/x)$$

so that we have, on writing the corresponding distance function at any point x

$$k(z) = \frac{\{g(z) - r(z)\}}{r(z)}$$

$$\frac{\{h(z) - r(z)\}}{r(z)} = \int_1^z f(x)\,k\{g(z/x)\}\,dx$$

$$+ \int_z^b f(x)\,k\{g(bz/x)\}\,dx \qquad (5.6\text{–}4)$$

If we assume that the distance of $g(x)$ from the reciprocal is some number $D\{g(z)\}$, then

$$|D\{h(z)\}| \leq D\{g(z)\}\left\{\int_1^z f(x)\,dx + \int_z^b f(x)\,dx\right\} \qquad (5.6\text{–}5)$$

$$\leq D\{g(z)\}$$

and therefore the distance after a multiplication is not greater than the distance from the smaller of the distances of the product terms from the reciprocal (since $f(x)$ and $g(y)$ play symmetric roles).

Since $g(z)$ and $r(z)$ are both probability distributions the difference

$$g(z) - r(z)$$

in the numerators of (5.6–3) must change sign at least once in the interval $(1, b)$ and there will *usually* be a lot of cancellation in the integrals where we have merely put bounds. Thus, without going into great detail, we see that a sequence of products, all drawn from a flat distribution will rapidly approach the reciprocal distribution.

Example 5.6–3　　*Approach from a Flat Distribution*

If we start with a flat distribution for both factors, that is

$$f(x) = g(x) = \frac{1}{b-1}$$

then we have for the distribution of the product from Example 5.6–1,

$$h(z) = \frac{[b \ln b - (b-1)\ln z]}{(b-1)^2}$$

If we continue we get the distance function for k factors from a flat distribution we get

Number of factors	Distance	% of original distance
1	1.558	100.0
2	0.3454	22.2
3	0.0980	6.29
4	0.0289	1.85

This shows that when the numbers are drawn from a flat distribution the product rapidly approaches the reciprocal distribution.

For division the results are much the same, see Exercises.

Example 5.6–4　　*A Computation of Products from the Flat Distribution*

To check the formulas and the theory we made three experiments. We divided the range into 9, 99, and 999 intervals, picked the midpoint numbers as representatives of the intervals, and formed all the possible products. We compared the results with the result of Example 5.6–1.

Table for 9 intervals

Digit	Number of cases	% of cases	Theoretical
1	19	23.457	24.135
2	16	19.753	18.321
3	11	13.580	14.545
4	9	11.111	11.738
5	7	8.642	9.501
6	7	8.642	7.640
7	3	3.704	6.047
8	6	7.407	4.655
9	3	3.704	3.418
total=	81		

Table for 99 intervals

Digit	Number of cases	% of cases	Theoretical
1	2364	24.120	24.135
2	1799	18.355	18.321
3	1426	14.546	14.545
4	1143	11.662	11.738
5	931	9.499	9.501
6	756	7.714	7.640
7	584	5.959	6.047
8	462	4.714	4.647
9	336	3.343	3.418
total=	9999		

Table for 999 intervals

Digit	Number of cases	% of total	Theoretical
1	240,890	24.1373	24.135
2	182,844	18.3210	18.321
3	145,181	14.5472	14.545
4	117,124	11.7359	11.738
5	94,816	9.5006	9.501
6	76,256	7.6409	7.640
7	60,321	6.4418	6.047
8	46,478	4.6471	4.655
9	34,091	3.4159	3.418
total=	999,999		

Thus we see that the granularity of the actual number system used in computers is successfully approximated by the continuous model of infinite precision - that the continuous model is robust with respect to the actual discrete computer model used in practice.

Exercises 5.6

5.6–1 Find the formula for the product of two numbers from a flat distribution, and then for four numbers.

5.6–2 The same for quotients.

5.6–3 Find the formula for the product of two numbers divided by a third.

5.6–4 Plot the difference computed in Exercise 5.6–3.

5.6–5 Plot the relative distance in Exercise 5.6–4.

5.6–6 Examine the rapidity of the approach in Exercise 5.6–3 from the flat to the limiting reciprocal distribution by trying the product of two such numbers.

5.6–7 Test Exercise 5.6–3 using 1 and 2 digit numbers.

5.6–8 Do a Monte Carlo test of Exercise 5.6–3.

5.7 Random Times

Apparently our instincts about time endow the flow of time with a bit more continuity than that of space, and we seem less likely to try to think of things happening in infinite time than on an infinite line or infinite plane. Furthermore, the Bernoulli type argument against the use of the expected value for money (Section 3.7) does not necessarily apply when we consider time. In estimating the average rate of failure of a machine the expected value is very useful and apparently is appropriate.

We first consider the following example where we pick a random time to observe.

Example 5.7–1 *Sinusoidal Motion*

Suppose a target is moving up and down in a sinusoid across our line of vision, and we want to hit it, but must aim and fire at a random time. We feel that it is reasonable to pick the random time in suitable half period π, which we can take from $-\pi/2$ to $\pi/2$ (this covers exactly once all values that can occur). We pick the motion as

$$y(t) = \sin t \qquad |t| \leq \pi/2$$

which clearly exhibits the assumption of the uniformity in time. We have by elementary calculus

$$t = \arcsin y$$

$$dt/dy = \frac{1}{\sqrt{1 - y^2}}$$

Now the probability of finding the target at a position between y and $y + \Delta y$ must be approximately

$$\Delta P = \frac{1}{\pi} \frac{1}{\sqrt{1 - y^2}} \Delta y$$

from which the probability density is

$$p(t) = \frac{1}{\pi \, |\cos t|}$$

We see that aiming at the midpoint of the interval (the expected value) is poor policy—we should pick one or the other end. Of course, since things in practice are finite in size, we are not dealing with exact points, so we need to pick our direction in a small angle from one of the extremes, but this detail does not alter the fact that the average (expected) position is not where to aim, but rather we should aim near one of the extremes.

Although we started out with a random point in time we were apparently not seriously bothered, in this case of a periodic motion where we picked a characteristic finite interval of behavior, but it needs careful thinking to convince yourself that we are not in the same position as picking a random point on an infinite line—both the periodicity, and, to a curious extent for many people, the use of time, seems to justify the process.

We believe that in this world many events occur approximately at random times, for example the decay of a radioactive atom, the time the next telephone call is placed on a central office, the time the next "job" is submitted to a computer, the time of the next death in a hospital, the time of the next failure of a machine (or one of its parts), etc. Each seems to be more or less random. We need to develop the mathematical tools for thinking about such things, including the probability density function $p(t)$.

We will use the standard approach of examining the finite case and then let the number of intervals approach infinity as the maximum length of an interval approaches zero—the standard method for setting up integrals.

Example 5.7–2   Random Events in Time

Consider, first, a fixed time interval $(0 \leq x \leq t)$. Let it be divided into n equal-sized subintervals each of length t/n. If each subinterval has the same probability of an event occuring, say p, and they are independent (Bernoulli), then the probability of exactly k events in the n intervals is

$$P_k(t) = b(k; n, p) = C(n, k)p^k(1 - p)^{n-k}$$

Now fix n and p. We next further subdivide each of the n intervals into m equal parts, and hence from the uniformity assumption we have the probability of an event in the new subintervals of length t/mn is p/m.

We have for the finer subintervals the same formula with new parameters

$$P_k(t) = C(nm, k)(p/m)^k(1 - p/m)^{mn-k}$$

It is now merely a matter of rearranging this expression before we take the limit as $m \to \infty$. We simply write out the terms in detail and reorganize the expression

$$P_k(t) = \{mn(mn - 1)(mn - 2) \cdots (mn - k + 1)/k!\}$$

$$\times (p/m)^k (1 - p/m)^{mn} (1 - p/m)^{-k}$$

$$= n(n - 1/m)(n - 2/m) \cdots$$

$$\times (n - (k - 1)/m)\{p^k/k!\}[(1 - p/m)^{-m/p}]^{-np}[1 - p/m]^{-k}$$

Now we let $m \rightarrow \infty$, remembering that k is fixed, and we get

$$P_k(t) \rightarrow n^k \left[p^k / k! \right] e^{-np} [1]$$

$$= \frac{(np)^k}{k!} e^{-np}$$

But recall that np is the expected number of successes in the original interval t. Hence if we now write **a** as the expected number of events in unit time we have

$$np = at$$

and hence

$$P_k(t) = \frac{(at)^k}{k!} e^{-at} \qquad (5.7\text{--}1)^*$$

as the limiting probability of k events in an interval of length t.

If we set the number of events in time t, $at = k$, we will get

$$P\{X = k\} = \frac{\lambda^k e^{-\lambda}}{k!}$$

The generating funtion of this *Poisson distribution* is (t in no longer time)

$$\text{G.F.} = \sum_0^\infty t^k \lambda^k e^{-\lambda} / k! = e^{(\lambda - 1)t}$$

To understand this formula, (and the fact that unusual events have to happen frequently according to this formula) we examine the probability of $0, 1, 2, \ldots$ occurrences in the time $t = 1/a$ (which is the time interval in which you expect a single event to occur). We have

$$\begin{aligned}
P_0(1/a) &= 1/e &&= 0.3679 \\
P_1(1/a) &= 1/e &&= 0.3679 \\
P_2(1/a) &= 1/2e &&= 0.1839 \\
P_3(1/a) &= 1/3!e &&= 0.0613 \qquad (5.7\text{--}2) \\
P_4(1/a) &= 1/4!e &&= 0.0153 \\
P_5(1/a) &= 1/5!e &&= 0.0031
\end{aligned}$$

$$\cdots \qquad \cdots$$

Thus, in an interval in which you expect to see a single event the actual probabilities of no event and of 1 event are the same, each about 37%. The probability of 2 events is only half that of no event, about 18%; and of 3 events 1/6 that of no event, about 6%; etc. In words, in an inteval in which the expected number of events is 1 you can expect 3 events about 6% of the time (in the time interval $t = 1/a$). The bunching of events is much more likely than most people expect.

We also see that the sum of all the values is exactly 1, since the system must produce one of the events (outcomes) $P_k(t)$ for each t. To demonstrate this mathematically we recall that the terms of the exponential series for e are simply $1/n!$

The model has the property (due to the independence assumed originally) that if we write $t = t' + t''$ we will have

$$P_0(t) = P_0(t')P_0(t'') \qquad\qquad (5.7\text{--}3)^*$$

Thus the probability of not seeing an event in time t is the product of not seeing it in time t' multiplied by the probability of not seeing in the next time interval t'' reaching up to t.

This is called "a constant failure rate"; what has happened up to the present time has no effect on what you will see in the future—but that was the independence assumption! From this comes the general rule for such situations—"If it is running then leave it alone!" You cannot improve things by tinkering. Indeed, it was the author's experience many years ago that the probability of failure in the electronic parts of a big computer was higher immediately after "preventive maintenance" than it was before! Currently preventive maintenance on electronic gear is usually limited to changing filters, and checking the mechanical parts. Of course mechanical wear and tear do not follow the constant failure rate that is typical of electronic gear (fairly accurately).

Example 5.7–3   *Mixtures*

We suppose: (1) that we have mixed chocolate chips into a mass of dough with a density of **a** per cookie (to be made), (2) that the presence of any chip in a cookie is independent of the presence of other chips, and (3) that the cutting up of the mass of dough into separate cookies is independent of the number of chips in the cookie—all large idealizations from reality. We can think of the cookies as being extruded and that the cookies are cut off when the *expected* number of chips is exactly **a.** Hence we have the above distribution. For example, if we expect an average of 4 chips per cookie, then the probabilities of k chips in a cookie are given by, (for $k = 0, 1, \ldots$)

TABLE 5.7–1

no chips	$\exp(-4)$	$= 0.01832$	$\sim 2\%$
1 chip	$4\exp(-4)$	$= 0.07326$	$\sim 7\%$
2 chips	$(4^2/2!)\exp(-4)$	$= 0.14653$	$\sim 15\%$
3 chips	$(4^3/3!)\exp(-4)$	$= 0.19537$	$\sim 20\%$
4 chips	$(4^4/4!)\exp(-4)$	$= 0.19537$	$\sim 20\%$
5 chips	$(4^5/5!)\exp(-4)$	$= 0.15629$	$\sim 15\%$
6 chips	$(4^6/6!)\exp(-4)$	$= 0.10420$	$\sim 10\%$
7 chips	$(4^7/7!)\exp(-4)$	$= 0.05954$	$\sim 6\%$
8 chips	$(4^8/8!)\exp(-4)$	$= 0.02977$	$\sim 3\%$
9 chips	$(4^9/9!)\exp(-4)$	$= 0.01323$	$\sim 1\%$
10 chips	$(4^{10}/10!)\exp(-4)$	$= 0.00529$	$\sim 0.5\%$
		Total	$= 99.5\%$

This table shows, for most people, that the deviations from the expected number are much larger than they had expected.

5.8   Dead Times

It is necessary to observe that sometimes the device that measures the events has a "dead time"; immediately after recording an event it cannot record a second event that comes too close after the first one. If the counting rate is comparatively slow with respect to the dead time then it is probably not worth making the corrections. But when you are pushing the recording equipment near its limit then the lost counts is a serious matter. In practice it is more difficult to find the realistic dead time than it is to allow for the multiple events that are recorded as single events. The ratio of single to double events, as given in the previous section, often gives you a reasonable first measure of the lost events.

5.9   Poisson Distributions in Time

Up to now the time t was fixed; now we look at things as a function of t. As we think of the probability density function we realize that we need to be careful. To get the notation into the standard form we need to replace k by n, and use $\Delta(at) = a\Delta t$. This produces an extra a in the probability density function

$$p_n(t) = a^{n+1} t^n \frac{e^{-at}}{n!} \qquad\qquad (5.9\text{–}1)^*$$

for $a > 0$ and all $t \geq 0$.

To check that we are right we compute the total probability for each state, that is we compute

$$J(n) = \int_0^\infty p_n(t)\, dt$$

It is convenient to change variables immediately to get rid of the letter **a** by setting

$$at = x \qquad \text{then} \qquad a\, dt = dx$$

The integral becomes

$$J(n) = \frac{1}{n!} \int_0^\infty x^n\, e^{-x}\, dx$$

Integration by parts, $U = x^n$, $dV = \exp(-x)\, dx$, gives, when the limits are substituted into the integrated part the reduction formula, for $n > 0$,

$$J(n) = \frac{1}{(n-1)!} \int_0^\infty x^{n-1} e^{-x}\, dx = J(n-1)$$

The case $n = 0$ gives, of course,

$$\int_0^\infty e^{-x}\, dx = 1$$

Hence

$$J(n) = 1 \qquad \text{for all } n.$$

and therefore each $p_n(t)$ is a probability distribution—for every n if you wait long enough you will surely see the n events.

Example 5.9–1 *Shape of the Probability of the State $p_n(t)$*

Except for the case $n = 0$ the distributions $p_n(t)$ have the value 0 at $t = 0$; for $n = 0$ the value at $t = 0$ is 1. We next seek the maximum value of $p_n(t)$ for $n > 0$, (since the case $n = 0$ is just the exponential). Differentiate the function and set the derivative equal to zero. Neglecting front constants, which do not enter into finding the location of the extremes, we have

$$nt^{n-1}e^{-at} - at^n e^{-at} = 0$$

The zeros occur at

$$n = at, \quad \text{and} \quad t^{n-1} = 0$$

The location of the maximum is at $t = n/a$, while the $n - 1$ values at $t = 0$ are the minima, $(n > 1)$.

The value of the distribution at the maximum is

$$p_n(n/a) = a^{n+1}\left\{\frac{n}{a}\right\}^n \frac{e^{-n}}{n!} = an^n \frac{e^{-n}}{n!}$$

To get an idea of this value we use the Stirling approximation for $n!$

$$p_n(n/a) = \frac{an^n e^{-n}}{n^n e^{-n}\sqrt{2\pi n}} = \frac{a}{\sqrt{2\pi n}}$$

To find the inflection points we need to find the second derivative and equate it to zero. We get for $(n > 1)$

$$n(n-1)t^{n-2}e^{-at} - 2nt^{n-1}a e^{-at} + t^n a^2 e^{-at} = 0$$

Hence we have (neglecting the roots at $t = 0$)

$$(at)^2 - 2n(at) + n(n-1) = 0$$

The solutions of this quadratic are, from the standard formula,

$$at = n \pm \sqrt{n^2 - n(n-1)} = n \pm \sqrt{n}$$

The positions of the inflection points are symmetricaly placed about the maximum, and we see that as n increases the maximum moves out proportional to n while the inflection points hug the mean like $\sqrt{n}$. *Relative* to the location of the maximum the width of the main peak gets narrower and narrower, like $1/\sqrt{n}$.

Exercises 5.9

    5.9-1   Find the mean and variance of the $p_n(t)$ distribution.

    5.9-2   Sketch these distributions for $n = 0, 1, 2$ and 3.

5.10 Queuing Theory

A queue is a common thing in our society, and the theory is highly developed. The basic model is that people, items for service, or whatever the input is, arrive according to some rule. In many situations, such as phone calls coming to a central office, the demand for service can be viewed as coming from a uniform random source of independent callers with a rate **r**, hence the probability density of the interarrival times is

$$p(t) = r\,e^{-rt}$$

The next stage to consider in the queuing process is the *server* that gives the service. In the simplest models, which are all we can consider here, the service may also be random with a mean service time of s. If the system is not going to ultimately be swamped then the *rate* of service must be faster than the arrival time rate, that is, $s > r$.

The general theory allows for many other rules for the arrival times and service times, but the the distributions we are assuming are quite common in practice—at least as first approximations.

When we think about the queue we see that occasionally there will be a burst of arrivals that temporarily exceeds the service capacity and the length of the queue will build up. In practice the queue may not be infinite and what to do with the overflow will differ in different situations. Similarly, for some purposes the order in which the items are served from the queue may matter; thus you may have first in first out (FIFO) or last in first out (LIFO), and there are many other *queue disciplines* as they are called.

We have, therefore, to think about the state of the queue (its length), and we are in the state diagram area (Section 4.4) with the interesting restriction (usually) that the transitions are only between adjacent states, one more customer arrives or else one more service is completed. And we also face the interesting fact that potentially there are an infinite number of states in the state diagram. Thus for being in state $P_0, P_1, P_2, \ldots, P_k, \ldots$ (with k items in the queue) we have the corresponding probabilities at time t of $P_0(t)$, $P_1(t)$, $P_2(t), \ldots, P_k(t), \ldots$ see Figure 5.10–1.

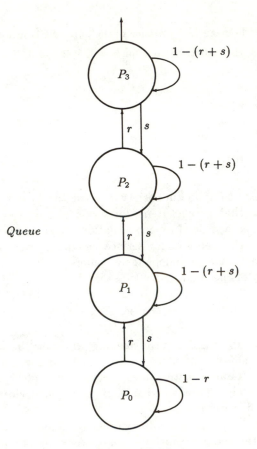

Queue

FIGURE 5.10–1

We now write the transition probability equations. The probability of the queue being in state k at time $t + \Delta t$ arises from (1) staying in the state, (2) coming from the state $k - 1$ by another arrival, or (3) coming from state $k + 1$ by the completion of a service. We neglect the chance of two or more such events in the small time Δt; ultimately we will take the limit as Δt approaches zero, and a double event is a higher order infinitessimal than a single event. We have

$$P_k(t + \Delta t) = \{1 - (r + s)\Delta t\}P_k(t) + \{r\Delta t\}P_{k-1}(t) + \{s\Delta t\}P_{k+1}(t)$$

where, of course, the state $P_{-1}(t) = 0$. We have, upon rearrangement,

$$\frac{P_k(t + \Delta t) - P_k(t)}{\Delta t} = rP_{k-1}(t) - (r + s)P_k(t) + sP_{k+1}(t)$$

In the limit as $\Delta t \to 0$ we get the differential equations

$$\frac{dP_k(t)}{dt} = rP_{k-1}(t) - (r+s)P_k(t) + sP_{k+1}(t)$$

If we start with an empty queue, then the initial conditions are:

$$P_0(0) = 1, \quad P_k(0) = 0, \quad k > 0.$$

To check these differential equations we add them and note that each term on the right occurs with a total coefficient (in the sum) of exactly 0, since the first few equations are clearly

$$\frac{dP_0(t)}{dt} = -rP_0(t) + sP_1(t)$$

$$\frac{dP_1(t)}{dt} = rP_0(t) - (r+s)P_1(t) + sP_2(t)$$

$$\frac{dP_2(t)}{dt} = rP_1(t) - (r+s)P_2(t) + sP_3(t)$$

$$\frac{dP_3(t)}{dt} = rP_2(t) - (r+s)P_3(t) + sP_4(t)$$

$$\cdots \qquad \cdots \qquad \cdots \qquad \cdots$$

Since the sum of the derivatives is zero the sum of the probabilities is a constant, and from the initial conditions this constant is 1; probability is conserved in the queue.

There is a known generating function for the solutions $P_k(t)$ which involves Bessel functions of pure imaginary argument but it is of little use except to get the mean and variance (which, since the Bessel function satisfies a linear second order differential equation, are easy to get from only the first derivative).

What is generally first wanted in queuing problems is the *equilibrium solution*, the solution that does not involve the exponentials which decay to zero as time approaches infinity. We will not prove that all the eigenvalues of the corresponding infinite matrix have negative real parts, so that they decay to zero, but it is intuitively clear that this must be so (since the probablities are non negative and total up to 1).

The equilibrium solution naturally has the time derivatives all equal to zero—hence at equilibrium we have the infinite system of simultaneous linear algebraic equations on the right hand sides to solve. They are almost triangular and that makes it a reasonable system to solve. We begin by *assuming* some unknown value for the equilibrium solution value P_0. From the the first equation we get

$$P_1 = \frac{r}{s} P_0$$

Using these two values we get from the next equation

$$P_2 = \frac{1}{s}[(r + s)P_1 - rP_0]$$

and some algebraic simplification gives

$$P_2 = \left(\frac{r}{s}\right)^2 P_0$$

It is an easy induction (with P_0 as a basis) to show that

$$P_n = \left(\frac{r}{s}\right)^n P_0$$

We did not know the value of P_0 with which we started, but merely assumed some value P_0. Using the fact that we know that the sum of all the probabilities must be exactly 1, we have

$$1 = \sum_{n=0}^{\infty} P_n = P_0 \sum_{0}^{\infty} \left(\frac{r}{s}\right)^n$$

The sum is a geometric progression with a common ratio (r/s) less than 1 so that the series converges). Hence we get

$$1 = P_0 \left[\frac{1}{1 - r/s}\right]$$

and

$$P_0 = 1 - r/s$$

Thus the equilibrium solution is

$$P_n = \left(\frac{r}{s}\right)^n (1 - r/s) = \left(\frac{r}{s}\right)^n - \left(\frac{r}{s}\right)^{n+1}$$

It is necessary to dwell on what this solution means. It does *not* mean that the queuing system settles down to a set of fixed values—no! The queue continues to change as customers enter the system and have their service completed. The formula means that the probabilities *you estimate* for the queue length at some distant future time t will settle down to these fixed values. Due to the randomness of the arrivals the actual queue length continues to fluctuate indefinitely. The equilibrium state gives you probability estimates in the long run, but as we said a number of times before what you "expect" to see and what you "actually" see are not the same thing. It is evident, however, that for many purposes the equilibrium solution is likely to be what you want, and you are not interested in the transient of the probability distributions which involves the exponential solution.

We began with a very simple model of a queue, but if we are willing to settle for the equilibrium solution then we are not afraid of much more complex situations. The actual time dependent solutions are not hard to find in this case.

Exercises 5.10

   5.10–1   If $r = ks(k < 1)$ compute P_n. Evaluate for $k = 1/2$ to $k = 9/10$.

   5.10–2   In Exercise 5.10–1 find the generating function of the queue length, and hence the mean and variance.

5.11   Birth and Death Systems

In birth and death systems (the name comes from assuming that items are born and die in various states and when this happens you migrate to the adjacent state, up for birth and down for death) we assume again that in the state diagram only adjacent states are connected, but this time we have that the probability of going to a state depends on the state that you are in, that is r and s are now dependent on the state you are in. In Figure 5.10–1 the transition probabilities from states to states acquire subscripts. By exactly the same reasoning as before we are led this time to the equations

$$\frac{dP_n(t)}{dt} = r_{n-1}P_{n-1}(t) - (r_n + s_n)P_n(t) + s_{n+1}P_{n+1}(t)$$

$$\frac{dP_0(t)}{dt} = -r_0 P_0(t) - + s_1 P_1(t)$$

   To get the equilibrium solutions we proceed as before; we first put the time derivatives equal to zero, assume a first value for P_0 state, and then solve the equations one at a time. We find

$$P_1 = \frac{r_0}{s_1} P_0$$

$$P_{n+1} = \frac{r_n}{s_{n+1}} P_n$$

From which we easily find by successive substitutions

$$P_n = \frac{r_0 r_1 \ldots r_{n-1}}{s_1 s_2 \ldots s_n} P_0$$

Our equation for the total probability is now

$$1 = P_0 \left[1 + \frac{r_0}{s_1} + \frac{r_0 r_1}{s_1 s_2} + \cdots + \frac{r_0 r_1 \ldots r_{n-1}}{s_1 s_2 \ldots s_n} + \cdots \right]$$

If and only if this series in the square brackets converges to some finite value S can we get the value of $P_0 = 1/S$, and hence get all the other values P_n. The divergence of the series implies that there is no equilibrium solution.

We have a general equilibrium solution, and if we want a more compact solution then we have to make some more assumptions (beyond the convergence) on the forms of the r_n and s_n. Whenever you can solve the infinite almost triangular system of algebraic linear equations in a neat form you can get the corresponding neat solution. We will not go into the various cases here as this book is not a course in birth and death processes nor in queuing theory. The purpose is only to show the range of useful problems that fall within the domain of simple probability problems and state diagrams.

Exercises 5.11

5.11–1 Show that for no deaths the solution is $p_n = \exp(-rt)(rt)^n/n!$

5.12 Summary

We did not lay down postulates and make deductions for the case of continuous probability, rather we proceeded sensibly and from a few simple Examples we saw the kinds of things we might want to assume.

We gave a number of Examples which show that it is non-trivial to decide which, if any, probability distribution to assume is uniform. The assignment of the probability distributions in the continuous case is often a serious problem (that tends to be glossed over in mathematics by making convenient assumptions), and clearly affect the results found after the more formal manipulations. The topic is discussed in much more detail in Chapters 6, 7 and 8.

When the independent variable is time then it is often easy and natural to assign the corresponding probabilities—typically a constant rate in time. Examples are bacterial growth, radioactive decay, failures in electronic systems, and service requirements.

6

Uniform Probability Assignments

Small Causes—Large Effects

6.1 Mechanical Probability

Now that we see the central role of the uniform distribution in probability problems we understand its importance and why we need to examine it more closely.

Based mainly on symmetry we began by assigning initial uniform probability distributions for events. If some obvious symmetry is lacking how can we justify assigning a uniform distribution? One answer given by many people, most notably by Poincaré, is based on the idea that even with a perfect (classical) mechanical situation the initial conditions are rarely perfectly known and often the result has a uniform distribution. (See also the modern theory of chaos.)

A simple example of randomness in the ideal physical world of classical mechanics is the thought experiment of dropping an ideal ball down a chute as consistently as possible. We imagine the ball falling onto a knife edge and then bouncing to one side or the other. We also imagine the result being tabulated and the ball returned to the device for another trial. Let us consider the results of many trials as a function of the displacement of the knife edge with respect to the center of the chute. When the knife edge is far to one side the ball will always fall to one side, and when it is far on the other it will fall on the other side. We are concerned with what happens when the knife is near the middle. As we imagine, in this ideal experiment, adjusting the equipment to get a 50–50 ratio for one side or the other we think that we will find positions which are close to this and where a slight movement one way or the other will change the ratio. We have, therefore, a conceptual mechanical device which can, apparently, produce a random behavior. Which side the ball will fall (on any particular trial) is not predictable, but the average behavior is reasonably well predicted; randomness in the ideal physical world. This result is based on *your* feeling of continuity, that as we make very small changes in the position of the knife edge the ratio of the number of events on the two sides cannot change suddenly from all on one side to all on the other, but must pass smoothly from one to the other. (In the extreme ideal world of exactness the exact middle position would have the ball bounce repeatedly on the knife edge and finally come to rest on it in a metastable state!)

We next consider the standard roulette wheel that is used in gambling. The wheel was apparently carefully designed to give uniform random results. We first make the assumption that the laws of classical mechanics are a perfect description of reality. The wheel is started rotating at a modest rate in one direction, and then a short time later a small ball is launched on the wheel going rapidly in the opposite direction. Due to the speed of the ball its centrifigual force makes the ball go promptly out to the outer containing rim of the wheel, where it continues to run around the track. As time goes on friction removes some of the speed from the ball (and a slight amount from the much more massive wheel) and the ball gradually approaches the inner track where there are 38 (or 37) equally sized holes. The moment the ball hits the edge of a hole it bounces out away from the line of holes (the wheel is rotating against the direction of the ball) and then comes down at some other place to hit another edge of a hole, etc., until it finally has acquired about the rotational rate of the wheel and comes to rest in one of the 38 identical holes.

Consider, now, the mechanics of the system. There is the speed of the wheel, the phase of the rotation (the relative position) of the wheel at the moment of launching the small ball, the three coordinates of the ball at the point of launch from the hand, the three corresponding velocities, two numbers to describe the axis of rotation of the ball, and finally the rate of rotation of the ball itself at the moment of launch. We have mentioned, in total, 11 different numbers that are needed (there may be more) to describe the trajectory of the ball. Thus for each possible launching of the ball, (in some suitable set of ranges that are possible), there is a point in an imagined 11 dimensional space. Nearby points in this space will have very different trajectories. There is a theorem in (ideal) mechanics which states that if the initial conditions of launch are varied continuously then the trajectory changes continuously. But as we imagine any one of the 11 parameters being changed very slightly we feel that the result will probably end up in another hole. Infinitesimal changes (almost in the mathematician's sense!) produce significantly different trajectories.

Now let us fix in our mind the terminal hole, and look back at the situation. Where in the 11 dimensional sample space do the balls come from which end up in the hole? The answer is that scattered throughout the whole of the 11 dimensional space of admissable initial conditions there are many very small regions that end there; each of these very small regions will have, however, quite different trajectories from other small regions.

No matter how we assign any reasonably smooth probability distribution over the launching space we *feel* that each hole will have almost the same probability of the ball ending there as any other hole. We do not bother to prove this mathematically, (but see the next section), we merely think of the almost trivial differences that will cause the ball to bounce at a slightly different angle, come off the rim at a slightly different place with respect to the position of the holes, etc. Indeed, we feel that a mote of dust could change the outcome, as well as a whiff of air, the temperature of the ball, and many other effects that we left out of the discussion. We simply do not expect that we could reproduce the same end result no matter how hard we tried, even in a laboratory, and we can not see any systematic effect preferring any one hole over another. Thus we come to the ideal uniform assignment of 1/38 for the probability of the ball ending up in any given hole.

There are three distinct things to be noted in this thought experiment. First, we have a situation in which the slightest differences, which are well below the level at which we can control things, lead to significantly different results. Second, that the trajectories are reduced modulo the rotation of the wheel, and the same end results come from all over the space of inital conditions. This type of reduction is a central feature—the trajectories *as trajectories* do not come together, rather as time goes on they spread out. Thus the initial assignment of a probability distribution in the launching space is still visible in the trajectory space—it is the reduction modulo the rotation of the wheel that brings the uniformity more or less independently of any reasonable initial probability distribution we might assume. Third, there is a final quantization—the ball must end up in a single hole—that further separates trajectories which started out very close to each other. In the current "chaos theory" it is constant folding and bifurcation that produces the unpredictable end result.

If we look further and consider what we believe we know about reality then we are even more convinced of the irreproducibility of results. The world is made out of molecules (so we believe most of the time) and these are in constant, rapid motion—yet it is the collision of the molecules of the ball with those of the edge of a hole that make the angle of rebound! A slight change of the angle due to chance positions of the molecules in an early collision might well alter the final position of the ball. We mentioned air drafts, motes of dust, and temperature (which could affect the size of the ball and hence its moment of inertia) among other things as other possible effects. If we go further and invoke quantum mechanics, which in its popular Copenhagen interpretation claims that the universe itself has a basic uncertainty, then we are even more convinced that it is *reasonable* to assign a uniform distribution to the final positions—*provided* upon examination we can detect no significant differences in the sizes or positions of the holes—no lack of symmetry there.

When we turn to the rolling of dice in gambling, then things are much the same, though perhaps not so safe. The standard requirement is that the dice hit the surface of the table before they bounce off the end wall (which has been made to have many different angles). The dice bounce off the wall and back onto the surface of the table, and finally end with one of the six faces up. Again we have the three features: (1) very slight differences in the initial conditions produce large differences in the trajectories, (2) there is the reduction modulo the rotations of the dice, and (3) there is the final quantization to a single face on the top side. It would take a very accurate machine to produce repeatable results, but this time we are not quite so sure as we were for the roulette wheel that it could not be done—we wonder to what extent humans can control the roll of the dice sufficiently well to significantly affect the outcomes of the dice.

Not that the design of the device itself could not affect the probabilities of the various faces turning up! Loaded dice are a familiar idea. And one can imagine a roulette wheel with a small permanent magnet under, say, the 00 hole, and the core of the ball being of a ferrous metal. And ferrous balls could be interchanged with non-ferrous at the whim of the person running the wheel. Or the holes could be carefully shaped to make some more likely to catch the terminal end of a trajectory than others, a single bar between two holes that is unusually high can affect the probabilites of falling in various holes. We are not concerned here with such designed cheating; we are concerned with the idea that a distribution which is very close to uniform can arise from a situation where: (1) very small differences in the initial conditions produce a very wide range of results, followed by (2) a reduction modulo something relatively small with respect to the range of trajectories that occur (a rotation of the wheel or the die), and then (3) a quantization into a few possible final states.

In shuffling cards the uniformity is much less secure and more easily controlled by practiced hands. Indeed, the common habit of the shuffler letting someone else "cut the deck" before dealing is a tacit admission of the possibility of controlling the dealt cards. There are also such matters as some cards having more ink on their faces than others, hence possibly different coefficients of slipperiness, etc.

It is not the aspect of cheating that we care about, rather it is the way situations can arise that call for the assignment of a uniform final distribution that is of interest. To repeat one more time, the esentials are: (1) small differences in the inital conditions give rise to large differences in the trajectories, (2) the reduction modulo something that is small with respect to the distribution of the endings of the trajectories viewed as total trajectories, and

(3) the final quantization to a definite state.

The matter may be summed up by the heading of this chapter:

Small Causes—Large Effects.

For further material see [E].

6.2 Experimental Results

In this section we will examine three simple distributions reduced modulo 1, which in the first and third cases is their variance. For example, consider a *normal distribution* with unit variance

$$p(x) = \frac{1}{\sqrt{2\pi}} e^{-x^2/2} \qquad (-\infty, \infty)$$

The reduction modulo 1 means that we sample the distribution at unit spacing and add all the values together for each possible phase shift. We will evaluate, therefore, (n is an integer and x is the phase shift displacement in a unit interval)

$$p(x) = \frac{1}{\sqrt{2\pi}} \sum_{n=-\infty}^{\infty} e^{-(x+n)^2/2}$$

The results, along with similar distributions, is given in the following table due to R. Pinkham, where x is the position in the unit interval of the new distribution after the modular reduction. The three columns contain the sum from all the corresponding values in the intervals that have been reduced modulo 1 of their distribution.

TABLE 6.2–1

x	*normal*	$(1/2)\exp\{-|x|\}$	$(1/2)\mathrm{sech}^2 x$
0.0	1.000 000 005	1.082	1.002
0.1	4	1.037	1.002
0.2	2	1.003	1.001
0.3	0.999 999 998	0.979	0.999
0.4	6	0.964	0.998
0.5	5	0.959	0.998
0.6	6	0.964	0.998
0.7	8	0.979	0.999
0.8	1.000 000 002	1.003	1.001
0.9	4	1.037	1.002
Sum of deviations 0		7	0

From the table we see that the reduction modulo 1 for the normal distribution is remarkably flat. Indeed in view of the Gregory integration formula [H, p.310, p.348], and the structure of the deviations from 1, we believe we are looking at roundoff and the detailed structure of the function. The rapid falloff of the higher derivatives as we approach infinity in the normal distribution case shows why it is so close to 1. In the other two cases the flatness is still remarkable, expecially for the two-sided exponential with its corner at the origin. The hyperbolic secant squared distribution is similar to the normal, but the flatness is not so spectacular because the higher derivatives do not fall off as rapidly as they do for the gaussian distribution.

There is a classic paper on this topic written by R. Pinkham [P] which shows in great generality that this reduction modulo something results in very flat distributions.

When we consider how three standard distributions, reduced by comparatively large unit reduction (as compared to the reductions in the sample space of the roulette wheel initial conditions), we see that we are remarkably confident that the resulting distribution for a roulette wheel will be very, very close to uniform in the theoretical model—in reality there will probably be larger faults with the physical realizations of the wheel and the uniform holes.

Exercises 6.2

6.2–1   Take your favorite continuous probability distribution and do a similar reduction modulo its variance.

6.2–2   Repeat 6.2–1 but use a spacing of half the variance.

6.2–3   Try the reduction using a spacing of twice the variance, and discuss the results of the three Exercises.

6.2–4   Explain the structure observed in Table 6.2–1.

6.2–5   Compute the corresponding table for the probability distribution $p(x) = \exp(-x)$, $(0 \leq x < \infty)$.

6.3 Mathematical Probability

There are a number of purely mathematical results that bear on this question of how uniform distributions can arise. The simplest result, and perhaps the most central, is a version of Weyl's theorem which states that if you: (1) take any irrational number **a**, (2) form all the multiples **na**, and (3) take the fractional part, then you will have a uniform distribution. These numbers may be written in the mathematical form

$$na - [na] = \text{fraction of } (na) = \text{frac}(na)$$

where [.] means the greatest integer in the argument.

Weyl's theorem has many important results. First, if $\log_{10} a$ is an irrational number then it follows that

$$a^n = 10^{(\log a)n} = 10^{[n \log a]} 10^{\text{frac}(n \log a)}$$

has the fractional part of the exponent uniformly distributed. The integer part of the exponent determines the placing of the decimal point while the fractional part determines the digits of the result. Hence we have the mantissa part uniformly distributed *up in the exponent*.

To show that this means that we have the reciprocal distribution (see Section 5.4) for the numbers themselves we argue as follows. First we have for a uniform distribution of a random variable *ln X* in the exponent

$$\Pr\{\log X < x\} = x \qquad (0 \le x \le 1)$$

But this is the same statement as

$$\Pr\{X < 10^x\} = x$$

Now write $10^x = y$, $(1 \le y \le 10)$, or $x = \log y$. We have

$$\Pr\{X < y\} = \log y = (ln\, y)/(ln\, 10)$$

To get the probability density we differentiate with respect to y

$$p(y) = 1/(y\, ln\, 10)$$

But this is exactly the reciprocal distribution! (Sections 5.4 to 5.6). And conversely, the reciprocal distribution implies that the distribution is uniform in the fractional part of the exponent.

Another example, in a sense, of Weyl's theorem is what is behind the standard random number generators. If we take an irrational number, the "seed number" **a** and multiply it (up in the exponent this is an addition) by a suitable "multiplier," then we will get from the product a fractional part that is a new seed number; and these repeated numbers (resembling the $na-[na]$) will be uniformly distributed in the interval $(0,1)$. Of course for finite precision computation it is necessary to apply some careful number theory arguments to ensure that with appropriate choices we will get a uniform distribution of $1/4$ of all the numbers that could possibly occur in the computer, [H, p.138]. Thus Weyl's theorem suggests that this method of random number generation is a reasonable one to try.

Not only do the mantissas of the powers of the typical numbers have the reciprocal distribution, so do many other sets of numbers. We will not prove it here but the leading digits of the factorials of the integers, the Bernoulli numbers, the binomial coefficients, and even the coefficients a_n of a Taylor series (provided there is only one singularity on the circle of convergence)

$$f(x) = \sum_{n=0}^{\infty} \frac{a_n(x - x_0)^n}{n!}$$

all satisfy the reciprocal distribution!

Exercises 6.3

6.3–1 Find the distribution of the fractional parts of the first 50 multiples of $\sqrt{2}$.

6.3–2 Similarly for $\sqrt{3}$.

6.3–3 Find the distribution of the leading digits of $n!$ for $n \leq 50$.

6.3–5 Examine a table of Fibonacci numbers and find the distribution of their leading digits.

6.3–6 Examine a table of Bernoulli numbers for the distribution of the leading digits.

6.3–7 Find the distribution of the leading digits of 2^n for $n = 1, 2, \ldots, 50$.

6.3–8 Write the Fibonacci numbers in binary form and find the distribution of the first three leading digits.

6.3–9 Find the distribution of the leading digits of 10^n when written in binary for $n = 1, 2, \ldots, 50$.

6.4 Logical Probability

Distribution of birthdays (see 1.9–6 for the original problem 1.9–7, 3.9–1, for robust versions and 3.9–3 for a variant) may depend on the phases of the moon as some maintain, but the earth and moon periods are irrationally related and the result is a rather uniform distribution when averaged over some years. But winter, spring, summer and fall are rationally related to the year. Only for events in the universe which do not seem to have any connection with the arbitrary rates of rotations of the earth about its axis and about the sun (which determines the length of a day and the year) do we feel that the day of the week on which, say, a super nova occurs is uniform (though, of course, the chance of its observation by humans may be affected by earthly details). Due to the seasons of the year we are not sure of the uniformity of birthdays throughout the year.

Thus we see that while we often argue for a kind of logical uniformity, it is not always safe reasoning. Still, very often we are unable to see any difference logically, and that brings up our earliest argument about the assigning of a measure of probability to interchangable events—if we do not assign the same measure then we are logically embarrassed. We do not have to surrender to logic—after all Emerson said, "A foolish consistancy is the hobgoblin of little minds."—but we are inclined to do so. Thus logical symmetry leads to a uniform distribution. However, later observed data that is not uniform may be an embarrassment!

We are actually discussing the famous *principle of indifference*, also called the principle of *insufficient reason*, but which should more accurately be called the *principle of consistancy*. It states that if there is no reason for assuming a difference then we assign a uniform probability.

But does this principle require perfect knowledge of *all* possible relevant facts (hardly attainable in reality), or does it merely require that we are ignorant of any lack of symmetry and have exercised "due prudence"? In the later case the idealist is inclined to be unhappy. This dilemma is a real difficulty and should not be ignored!

Philosophers, and some mathematicians, have had great fun mocking this principle of consistancy—but almost always without regard to what it says! It says that if you have *absolutely no reason* for preferring A to B, for example, then it is reasonable (unless you want to be accused of inconsistancy) to assign equal probabilities to each event. It is a hard, exacting principle; *there must be no reasons that you are aware of* for any difference. If there are slight differences, as there clearly are with regard to the six faces of a die, then you have to decide on an intuitive basis whether to go with the idealized die in your predictions or else to try to measure, in some fashion, the effects of six concave holes on one face with the opposite face having only one hole, (thus apparently moving the center of gravity of the die slightly from the exact center). Similarly for the other pairs of opposite faces. You have then to carry out the computations with the slight biases. The fact that we play so many games using dice suggests that practice finds the effects of these biases to be very slight. (See the next Section and Example 7.5–2.)

This principle of indifference (or consistancy) is very deep in our mental makeup, and the argument based on consistancy is very compelling; if you can see no relevant difference (that is, if there appears to be perfect symmetry in some sense) then any assignment other than equal probabilities will get you into a contradiction. Of course it depends on your seeing the symmetry. If you were fooled then you have made a mistake. If another person does not see your symmetry, but possibly some other symmetry, then your and their assignments of probability will differ. We cannot get perfect objectivity, we can only isolate it so that it can be examined closely. It can also be said that if the cost of an error is low we do not look carefully; if the cost is high we should.

6.5   Robustness

Since we have been arguing that it is reasonable to assume that many distributions are uniform, we need to look at the question of how the answer changes when there are perturbations from the flat distribution. This is a natural question to ask in any application of mathematical results to the real world where the exact hypotheses are seldom realized. We need to examine how "robust" the answers are. We have, in fact, already done a number of such examples (3.2–2, 3.4–2, 3.5–3, Sections 3.9, and 4.4–2), and we find that when the biases from a uniform distribution are small then their effects are generally quadratic in the biases.

This is a very general principle. When you are at an extreme, as is the uniform distribution (see the next Chapter) then small deviations in the independent variables produce only quadratic effects in the performance (as one sees from any elementary calculus course). Note that we have not defined "small" since it depends on the particular situation. Normally the extreme is relatively broad in character and hence the result is "robust." Once in awhile there is a very narrow extreme, and in such cases it is seldom practical engineering to use such designs; slight changes in various parameters will greatly affect the performance. In probability theory the sum of the probabilities of the deviations must be zero, hence there will often be compensating effects which tend to make the problem robust with respect to small deviations from the uniform distribution.

In Section 3.9 we examined the problem of robustness in two cases, the robust birthday problem and the robust elevator problem. In the first case we used the Taylor expansion about the symmetric point, found the constant term to be the symmetric solution, the first order terms, being at a local maximum always cancelled out, and the second order terms in the deviations from uniform gave a convenient formula for the perturbations about the uniform solution.

In other problems of the same general form we will often find that the second order derivatives with respect to the same variable are not all zero. By symmetry they will all, however, have the same value B for these at the symmetric point of evaluation, and the same value C for the cross derivatives. Again, expanding about the symmetric point $1/n$, and using the relations

$$\sum_{k=1}^{n} e(k) = 0$$

$$\left\{ \sum_{k=1}^{n} e(k) \right\}^2 = 0$$

we get

$$\sum_{k=1}^{n} e^2(k) = - \sum_{j \neq k} e(j)e(k)$$

Thus we have the useful formula for the expansion of the function $Q\{p(k)\}$

$$Q\{p(k)\} = Q(1/n) + (1/2)(B - C)n\,\mathrm{VAR}\{e(k)\} + \cdots \qquad (6.5\text{--}1)$$

A large class of uniform distribution problems can be solved in this form—we have only to find the common value of all the second order (same) derivatives, and the common value of all the cross derivatives, the numbers B and C, and plug these into this general formula (6.5–1).

We have sketched out how to cope with the robustness of the answer in a class of probability problems, but of course not for all probability problems. We will later examine further cases of robustness; we have already examined robustness in Example 3.3–2 where we asked for solutions as a function of p when the original problem assumed a definite p value.

If we cannot handle robustness in probability theory then we will always have questions of safety in applications of the results. Thus robustness is a central, much neglected, topic in probability problems.

6.6 Natural Variables

We have been arguing that in many situations it is natural to expect that the unknown distribution will be uniform. An argument often given against this assumption is that if the variable x is uniform $(0 < x < 1)$ then the variable x^2, or the variable $\sqrt{x}$, cannot be uniform. When the random uniform distributions arise from a modular reduction then this fact removes the relevance of the above observation; the variable that is so reduced is the "natural variable" to assume is uniform.

It is not always easy to decide just how the reduction modulo some constant is done before you get to see the random variable, but in a sense that is what is necessary to decide before assigning the uniform distribution (see Bertrand's paradox, Example 5.3–1). Unfortunately statistics will tell you almost nothing about what are the "natural variables" of the problem, although multidimensional scaling and other similar statistical tools exist to help in this matter. We have seen that the mantissas of natually occuring numbers are uniform in the fractional part of the exponent, and hence from the mathematical identity

$$e^{ln\ x} = x$$

it is the log of the mantissas of the numbers that arise in computing that is the natural variable for floating point arithmetic.

In physics, and other fields, experience has shown that the "natural variables" are usually the ones which enter into nice mathematical formulas. For example, Galileo first thought that the velocity of a falling body would be simply related to the distance covered, but found that it was the time elapsed that gave the simple mathematical formula.

There is no easy answer to the question, but the general indictment of the principle of indifference for assigning a uniform distribution is unjustified, *when* one can see that there is some modular reduction which is small with respect to the spread of the underlying variable. (See again Section 6.2).

This section on "natural variables" is of necessity intuitive, but the author believes that it tends to justify the frequent use of the uniform distribution when little else is known beyond some modular reduction before you see the results.

6.7 Summary

The initial assignment of the probabilities of the events in the sample space is fundamental. As Bertrand's paradox (and others) show, this is a serious matter since it affects the results obtained, and hence any future action taken. Unfortunately, this point is too often glossed over, and we have perhaps belabored it more than necessary [S].

In this Chapter we gave a number of arguments why the uniform distribution often occurs. The most widely used are: (1) the mechanical uncertainty ("small causes—large effects" which are then reduced modulo something, and often further quantized); and (2) the principle of consistancy (indifference). Unfortunately the latter depends on the way we look at the situation, not necessarily on reality!

The above two reasons apply quite well to the traditional gambling devices—roulette wheels, dice, and cards for example—but for other applications they are often much less reliable when applied. Most text books in probability simply assume a uniform distribution for the problems they pose; Nature sometimes is not so kind!

In the next two Chapters we will further pursue this central problem of the initial assignment of probabilities to the sample space, using somewhat more sophisticated arguments.

7

Maximum Entropy

"There is no doubt that in the years to come the study of entropy will become a permanent part of probability theory;..."

A. I. Khinchin (in 1933) [K,p.2]

7.1 What is Entropy?

The main purpose for studying entropy is that it provides a new tool for assigning initial probability distributions. The method is consistant with our symmetry approach but also includes nonuniform distributions. Thus it is a significant extension of the principle of symmetry.

Just as the word "probability" is used in many different senses, the word "entropy" also has several, perhaps incompatible, meanings. We therefore need to examine these different usages closely since they are the source of a great deal of confusion—especially as most people have only a slight idea of what "entropy" means. We will examine four different meanings that are the more common usages of the word. Since this is not a book on thermodynamics or statistical mechanics the treatment must, of necessity, be fairly brief and superficial in the thermodynamically related cases.

The word *entropy* apparently arose first in classical thermodymanics. Classical thermodynamics treats state variables that pertain to the whole system, such as pressure, volume and temperature of a gas. Besides the directly measurable variables there are a number of internal states that must be inferred, such as *entropy* and *enthalpy*. These arise from the imposition of the law of the conservation of energy and describe the internal state of the system. A mathematical equation that arises constantly is

$$dH = dQ/T$$

where dH is the change (in the sense of engineers and physicists) in entropy, T refers to the absolute temperature, and dQ is the quantity of heat transferred at temperature T. From the assumed second law of thermodynamics (that heat does not flow spontaneously from cold to hot) it is asserted that this quantity can only increase or stay the same through any changes in a closed system. Again, classical thermodynamics refers to *global properties* of the system and makes no assumptions of the detailed micro structure of the material involve.

Classical *statistical mechanics* tries to model the detailed structure of the materials, and from this model to predict the rules of classical thermodynamics. This is "micro statistics" in place of "macro statistics" of classical thermodynamics. Crudely speaking, matter is assumed to be little hard, perfectly elastic, balls in constant motion, and this motion of the molecules bouncing against the walls of the container produces the pressure of the gas, for example. Even a small amount of gas will have an enormous number N of particles, and we imagine a *phase space* whose coordinates are the position and velocity of each particle. This phase space is a subregion of a $6N$ dimensional space, *and* in the absence of any possibility of making the requisite measurements, it is *assumed* that for a *fixed energy* every small region in the phase space has the same probability as any other (compare the Poincaré theory of small causes, large effects). Then average values over the phase space are computed. Boltzmann broke up the phase space into a large number of very small regions and found that the quantity

$$H = k \log \frac{1}{P}$$

seemed to play a role similar to the classical concept of entropy (k is Boltzmann's constant and P is the probability associated with any one of the equally likely small regions in the phase space with the given energy.

Boltzmann attempted to prove the famous Boltzmann's *"H theorem"* which states that with increasing time H could not decrease but would generally increase. It was immediately pointed out by Loschmidt that since Boltzmann began with Newtonian mechanics, in which all the equations of motion that he used are reversible, then it is impossible from this *alone* to deduce irreversibility. A lot of words have been said about the probabilities of the states when all the velocities of the particles are reversed (to make things go backwards in time), and it is usually claimed that these states are much less likely to occur than those in the original direction—though so far as the author knows there never has been any careful mathematical proof, only the argument that this must be so in order for the theorem they wanted to be true! Kac gave an interesting simple counter example to the proof using only a finite system that was clearly periodic and reversible. (See Wa, pp.396–404)

This, then, is the second entropy that is widely used, and is thought to be connected with the "mixed-up-ness" of the system.

Gibbs, in trying to deal with systems which *did not have a fixed energy* introduced the "grand cannonical ensemble" which is really *an ensemble of the phase spaces* of different energies of Boltzmann. In this situation he deduced an entropy having the form

$$H = \sum_i p(i) ln \frac{1}{p(i)}$$

where the $p(i)$ are the probabilites of the phase space ensembles. This, then, is still a third kind of physical entropy.

Shannon while creating information theory found an expression of exactly the same form, and called it *entropy*, apparently on the advice of von Neumann, ("You should call it 'entropy' for two reasons: first, the function is already in use in thermodynamics under that name; second, and more importantly, most people don't know what entropy really is, and if you use the word 'entropy' in an argument you will win every time!") [L–T, p.2]

But the identity of mathematical form does not imply identity of meaning. For example, Galileo found that falling bodies obey the formula

$$s = \frac{g}{2} t^2$$

and Einstein found the famous equation

$$E = mc^2$$

but few people believe that the two equations have much in common except that both are parabolas. As the derivation of Shannon's entropy in the next section will show, the variables in his equation bear little resemblence to those of Gibbs or Boltzmann. However, from Shannon's entropy Tribus has derived all the basic laws of thermodynamics [L–T] (and has been ignored by most thermodynamicists).

While the identity of form cannot prove the identity of meaning of the symbols, it does mean that the same formal mathematics can be used in both areas. Hence we can use many of the formal mathematical results and methods, but we must be careful in the interpretations of the results.

7.2 Shannon's Entropy

Let X be a discrete random variable with probabilities $p(i)$, $(i = 1, 2, \ldots, n)$. We shall show in a few moments that when dealing with "information" the standard assignment for *the amount of information*, $I\{p(i)\}$, to be associated with an outcome of probability $p(i)$, is measured by the "surprise," the likelihood of the event, or the information gained, which is

$$I\{p(i)\} = \log \frac{1}{p(i)} = -\log p(i)$$

In the certain event $p = 1$, and no new information is obtained when it happens.

The result of this assignment is that the *expected value* of the information $I(X)$ for a random variable X is given by the formula,

$$H(X) = E\{I(X)\} = \sum_{i=1}^{n} p(i) \log \frac{1}{p(i)} \tag{7.2-1}$$

In *information theory* this is called the *entropy*. It is usually labeled $H(X)$, or sometimes $H(p)$.

Information theory usually uses the base 2 for the logarithm system, which we do in this chapter *unless* the contrary is explicitly stated. Logs to the base e will be denoted, as usual, by *ln*.

Kullback [Ku] has written a book (1959) with the title, *Information Theory and Statistics*, which shows many of the connections between the two fields.

This assignment of values to a random variable X is one of a number of different *intrinsic assignments of values* to the outcomes of a trial of a random variable, "intrinsic" meaning that the values come *directly* from the probabilities and have no *extrinsic* source. The general case of intrinsic assignments is the assignment of the value of the random variable as some function $f(p(i))$. In contrast, dealing with the time to first failure for a run of n trials gives an *extrinsic* but *natural* assignment of the index n as the values for the possible outcomes of probabilities qp^{n-1}, $(n = 1, 2, \ldots)$.

The entropy can also be viewed in a different way; since

$$p(i) \log \frac{1}{p(i)} = \log \left\{ \frac{1}{p(i)} \right\}^{p(i)}$$

we have

$$H(X) = -\log \prod_{i=1}^{n} p(i)^{p(i)}$$

and taking antilogs of both sides we have

$$2^{-H(X)} = \prod_{i=1}^{n} p(i)^{p(i)} \qquad (7.2\text{--}2)$$

which resembles a *weighted geometric mean* of the probabilities. Because of this logarithmic structure Shannon's entropy has many of the properties of both the arithmetic and geometric means.

Example 7.2–1   *An Entropy Computation*

Given a random variable X with the probability distribution $p(1) = 1/2$, $p(2) = 1/4$, $p(3) = 1/8$, and $p(4) = 1/8$, the entropy of the distribution is

$$H(X) = (1/2)\log 2 + (1/4)\log 4 + (1/8)\log 8 + (1/8)\log 8$$
$$= (1/2)(1) + (1/4)(2) + (1/8)(3) + (1/8)(3)$$
$$= 1/2 + 1/2 + 3/8 + 3/8 = 7/4$$

From *any* finite probability distribution we obtain the entropy as a single number—much as the mean is a single number derived from a distribution.

The entropy function measures the amount of *information* contained in the distribution, and is sometimes called the *negentropy* since it is thought to be the negative of the usual entropy which in turn is supposed to measure the "mixed-up-ness" of a distribution. We have already exposed the weakness of the analogy of Shannon's entropy with the entropies of physics, but see [T],[L–T]. For biology see [Ga].

Since we will be using Shannon's entropy extensively we need to make some reasonable derivation of his entropy function. Let there be two independent random variables X and Y with outcomes x_i and y_j (say the roll of a die and the toss of a coin). How much information is contained in the observed outcome $x_i y_j$? If $I(.)$ is the measure of the amount of information then we *believe* that for *independent random variables* the total information is the sum of the information contained in the individual outcomes, that is

$$I(x_i y_j) = I(x_i) + I(y_j) \qquad (7.2\text{--}3)$$

Think this over carefully before accepting it as fitting *your* views of information.

Equation (7.2–3) is the standard Cauchy functional equation when we *also* suppose that the measure of information $I(p)$ is a non-negative continuous function of its argument.

To study the solution of Cauchy's equation we drop the subscripts. Consider the *functional equation* (defines a function rather than a number)

$$f(xy) = f(x) + f(y) \qquad (7.2\text{--}4)$$

where $f(x)$ is a continuous nonnegative function of x. Suppose that $y = x$, then we have from equation (7.2–4)

$$f(x^2) = 2f(x)$$

Next, let $y = x^2$. Then from equation (7.2–4) we get

$$f(x^3) = 3f(x)$$

and in general, by induction, we have for all integers $n > 0$,

$$f(x^n) = nf(x)$$

Now if we write $x^n = z$ we have

$$f(z) = nf(z^{1/n})$$

or rewriting this

$$f(z^{1/n}) = (1/n)f(z)$$

Proceeding as before, we can show that for all rational numbers p/q

$$f(z^{p/q}) = (p/q)f(z)$$

By the continuity assumption on the solution of the equation since the equality holds for all rational numbers it holds for all real numbers, $x > 0$ and y

$$f(x^y) = yf(x) \tag{7.2–5}$$

The relationship (7.2–5) suggests that the solution is $f(x) = \log x$ to some base. To prove that this solution is unique to within a constant multiplier (effectively the base chosen for the logs) we assume that there is a second solution g(x), and then consider the difference

$$f(x) - kg(x) = \log_2 x - kg(x)$$

where, of course from (7.2–5) $g(x^y) = yg(x)$. We have for the difference

$$\log x - kg(x) = (1/y)\{\log x^y - kg(x^y)\}$$

We fix in our minds some $x = x_0$, (not 0 or 1), and choose a particular k so that at $x = x_0$ the left hand side is zero (so that the scaled second solution has a common point with the log solution at $x = x_0$). Hence we set

$$k = \frac{\log x_0}{g(x_0)}$$

We have on the right

$$\log x_0^y - kg(x_0^y) = 0$$

Now for any z other than 0 and 1 there is a y such that

$$z = x_0^y \quad \text{namely} \quad y = \frac{\log z}{\log x_0}$$

and we have

$$\log z = kg(z) = g(z)\frac{\log x_0}{g(x_0)}$$

Therefore any other solution $g(z)$ is proportional the $\log z$; and the solutions differ only with respect to the logarithm base used.

It is conventional to use, in information theory, the base 2 so that a simple binary choice has unit information; it is only a difference in the units used to measure information that is being determined by the choice of base 2. The standard equation

$$\log x = \frac{\log_b x}{\log_b 2} = (\log_2 b)(\log_b x)$$

means that every log system is proportional to every other, and the conversion from one to another is simple.

Since by assumption $I(p) \geq 0$ it is necessary to use a negative value for k so we finally choose our solution of equation (7.2–3) as

$$I(p) = \log\frac{1}{p} = -\log p \qquad (7.2\text{–}6)$$

Now that we have the measure of information for any outcome of probability p, we examine it to see its reasonableness. The less likely the event is to happen the more the *information*, the more the *surprise*, the greater the reduction in *uncertainty*. The certain event $p = 1$ has no information, while the impossible event has infinite information, but it never occurs! See below, equation (7.2–8).

The *average information* contained in a distribution $p(i)$ is clearly the weighted average of $\log\{1/p(i)\}$ (the expected value). From (7.2–6) we have

$$H(X) = \sum_{i=1}^{n} p(i)\log\frac{1}{p(i)}$$

$$(7.2\text{–}7)$$

$$= -\sum_{i=1}^{n} p(i)\log p(i)$$

which is *the entropy* of the distribution. The entropy is measured in units of bits per trial.

When we have a binary choice then we often write the argument as p (or q)

$$H(p) = p\log\frac{1}{p} + q\log\frac{1}{q} = H(q)$$

as as matter of convenience since the entropy then depends on the single number p (or q).

We need to examine the shape (graph) of the typical term of the entropy function as a function of the probability p. There is clearly trouble at $p = 0$. To evaluate this indeterminate form we first shift to logs to the base e and later rescale.

$$\lim_{p\to 0} p\,ln\frac{1}{p} = \lim_{p\to 0}\left[\frac{ln(1/p)}{1/p}\right]$$

We can now apply l'Hopital's rule,

$$\lim_{p\to 0}\left[\frac{-1/p}{-1/p^2}\right] = \lim_{p\to 0} p = 0 \tag{7.2–8}$$

and we see that at $p = 0$ the term $p\,ln(1/p) = 0$. Thus while the impossible event has infinite information the limit as p approaches 0 of the probability of its occurrence *multiplied* by the amount of information is still 0. We also see that the entropy of a distribution with one $p(i) = 1$ is 0; the outcome is certain and there is no information gained.

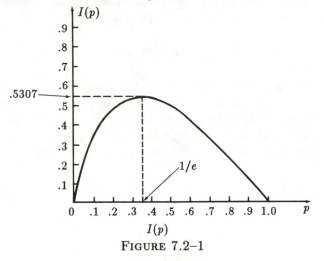

FIGURE 7.2–1

For the slope of the curve $y = p\,ln(1/p)$ we have

$$y' = -1 - ln\,p$$

and it is infinite at $p = 0$. The maximum of the curve occurs at $p = 1/e$. Thus, rescaling to get to logs base 2, we have the Figure 7.2–1.

In coding theory it can be shown that the entropy of a distribution provides the lower bound on the average length of any possible encoding of a uniquely decodable code for which one symbol goes into one symbol. It also arises in an important binomial coefficient inequality

$$\sum_{k=0}^{an} C(n, k) \leq 2^{nH(a)} \tag{7.2-9}$$

where $a < 1/2$ and *provided* n is made sufficiently large. Thus the entropy function plays an important role in both information theory and coding theory.

Example 7.2–2   *The Entropy of Uniform Distributions*

How much information (on the average) is obtained from the roll of a die? The entropy is

$$H(X) = \sum_{i=1}^{6} \frac{1}{6} \log 6$$

$$= \log 6 = 2.584962\ldots \text{ bits/roll}$$

Similarly, for any uniform choice from n things the entropy is

$$H(X) = \log n \quad \text{bits/trial}$$

which gives a reasonable answer. In the cases where $n = 2^k$ you get exactly k bits of information per trial.

Exercises 7.2

7.2–1   Show that (7.2–9) holds at $a = 1/2$.

7.2–2   Compute the entropy of the distribution $p(i) = 1/8$, $(i = 1, 2, \ldots, 8)$.

7.2–3   Compute the entropy of $p(i) = 1/3, 1/3, 1/9, 1/9, 1/9$.

7.2–4   Show that $\int_0^1 I(p)\, dp = 0.3606\ldots$.

7.2–5   For the probabilities $1/3, 1/4, 1/6, 1/9, 1/12, 1/18$ show that the entropy is $8/9 + (11/12) \log 3 \sim 2.34177$.

7.2–6   Let $S_N = \sum_{n=1}^{N} 1/n$. Then Zipf's distribution has the probabilities $p_n = (1/n)/S_N$. If $T_N = \sum_{n=1}^{N} (\log n)/n$ show that the entropy of Zipf's distribution is $H(N) = T_N/S_N + \log S_N$.

7.2–7   Using Appendix A.1 estimate the entropy of Zipf's distribution.

7.3 Some Mathematical Properties of the Entropy Function

We now investigate this entropy function of a distribution as a mathematical function *independent* of any meaning that may be assigned to the variables.

Example 7.3–1   *The log Inequality*

The first result we need is the inequality

$$ln\ x \leq 1 - x \qquad\qquad (7.3\text{–}1)$$

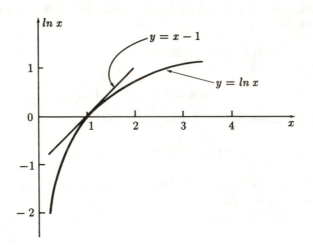

FIGURE 7.3–1

This is easily seen from Figure 7.3–1 where we fit the tangent line to the curve at $x = 1$. The derivative of $ln\ x$ is $1/x$, and at $x = 1$ the slope is 1. The tangent line at the point (1,0) is

$$y - 0 = 1(x - 1)$$

or

$$y = x - 1$$

Since $y = ln\ x$ is convex downward ($y'' = -1/x^2 < 0$) the inequality holds for all values of $x > 0$, and the equality holds *only* at $x = 1$. Compare with Appendix 1.B.

Example 7.3–2   *Gibbs' Inequality*

We now consider two random variables X and Y with finite distributions $p_X(i)$ and $p_Y(i)$, both distributions with the sum of their values equal to 1. Starting with (using natural logs again)

$$\sum_{i=1}^{n} p_X(i) \ln \frac{p_Y(i)}{p_X(i)}$$

we apply the log inequality (7.3–1) to each log term to get

$$\sum_{i=1}^{n} p_X(i) \ln \frac{p_Y(i)}{p_X(i)} \leq \sum_{i=1}^{n} p_X(i) \left\{ 1 - \frac{p_Y(i)}{p_X(i)} \right\}$$

$$\leq \sum_{i=1}^{n} p_X(i) - \sum_{i=1}^{n} p_Y(i) = 1 - 1 = 0$$

The equality holds only when each $p_X(i) = p_Y(i)$. Since the right hand side is 0 we can multiply by any positive constant we want and thus, if we wish, convert to base 2. This is the important *Gibbs' inequality*

$$\sum_{i=1}^{n} p_X(i) \log \frac{p_Y(i)}{p_X(i)} \leq 0 \qquad (7.3\text{–}2)$$

The equality holds only when $p_X(i) = p_Y(i)$ *for all i.*

Example 7.3–3   *The Entropy of Independent Random Variables*

Suppose we have two independent random variables X and Y and ask for the entropy of their product. We have

$$H(XY) = \sum_{i}^{j} p_X(x_i) p_Y(y_j) \log \frac{1}{p_X(x_i) p_Y(y_j)}$$

$$= \sum_{i}^{j} p_X(x_i) p_Y(y_j) \log \frac{1}{p_X(x_i)} + \log \frac{1}{p_Y(y_j)}$$

When we sum over the variable not in the log term we get a factor of 1, and when we sum over the variable in the log term we get the corresponding entropy; hence

$$H(XY) = H(X) + H(Y) \qquad (7.3\text{–}3)$$

When we compare this with (2.5–4)

$$E\{X + Y\} = E\{X\} + E\{Y\}$$

we see that the entropy of the product of *independent* random variables satisfies the same equation as the expectation on sums (independent or not), and partly explains the usefulness of the entropy function.

Suppose next that we have n independent random variables X_i, $(i = 1, \ldots, n)$, and consider the entropy of the compound event $S = X_1 X_2 \ldots X_n$. We have, by similar reasoning,

$$H(S) = H(X_1) + H(X_2) + \cdots + H(X_n) \qquad (7.3\text{--}4)$$

Thus n repeated trials of the same event X gives the entropy

$$H(S) = nH(X) \qquad (7.3\text{--}5)$$

This is what one would expect when you examine the original definition of information carefully.

Example 7.3–4   *The Entropy Decreases When You Combine Items*

We want to prove that if we combine any two items of a distribution then the entropy decreases, that we lose information. Let the two items have probabilities $p(i)$ and $p(j)$. Since probabilities are non-negative it follows that

$$p(i) \leq p(i) + p(j)$$
$$\log p(i) \leq \log\{p(i) + p(j)\}$$
$$\log 1/p(i) \geq \log \frac{1}{p(i) + p(j)}$$

Now: (1) multiply the last equation by p(i), (2) interchange i and j, and (3) add the two equations to get

$$p(i) \log \frac{1}{p(i)} + p(j) \log \frac{1}{p(j)} \geq \{p(i) + p(j)\} \log \frac{1}{p(i) + p(j)}$$

and we have our result which agrees with our understanding of Shannon's entropy: combining items loses information.

Exercises 7.3

7.3–1 What is the amount of information in drawing a card from a deck? Ans. $\log 52 = 5.7004397\ldots$ bits/draw.

7.3–2 What is the amount of information in drawing k cards without replacement from a deck?

7.3–3 Compute the entropy of the distribution $p_1 = 1/2$, $p_2 = 1/4$, $p_3 = 1/8$, $p_4 = 1/16$, $p_5 = 1/16$.

7.3–4 Write a symmetric form of Gibbs' inequality.

7.3–5 Show that when combining two items having the same probability $p(i)$ you lose $-2p(i) \ln 2$ of information.

7.3–6 If you have a uniform distribution of $2n$ items and combine adjacent terms thus halving the number of items, what is the information loss?

7.3–7 What is the information in the toss of a coin and the roll of a die?

7.3–8 What is the information from the roll of n independent dice?

7.3–9 Show that the information from the toss of n independent coins $= n$.

7.4 Some Simple Applications

We now show that Shannon's definition of entropy has a number of other reasonable consequences that justify our special attention to this function.

Example 7.4–1   *Maximum Entropy Distributions*

We want to show that the distribution with the maximum entropy is the flat distribution. We start with Gibbs' inequality, and assign the $p_Y(i) = 1/n$ (where n is the number of items in the distribution). We have

$$\sum_i p_X(i) \log \frac{1}{n\, p_X(i)} \le 0$$

Write the log term of the product as the sum of logs and transpose the terms in n. We have

$$H(X) = \sum_i p_X(i) \log \frac{1}{p_X(i)} \le \sum_i p_X(i) \log n = \log n$$

or

$$H(X) \le \log n$$

and we know that the Gibbs equality will hold when and only when all the $p_X(i)$ are equal to the $p_Y(i)$, that is $p_X(i) = 1/n$, which is the uniform distribution. When the entropy is maximum then the distribution is uniform and

$$\max\{H(X)\} = \log n \tag{7.4--1}$$

This includes the converse of Example 7.2--2.

For a finite discrete uniform distribution the entropy is a maximum; and conversely if the entropy is maximum the distribution is uniform. This provides some further justification for the rule that in the absence of any other information you assign a uniform probability (for discrete finite distributions). Any other finite distribution assumes some constraints, hence the uniform distribution is the most unconstrained finite distribution possible. See Chapter 6.

This definition of the measure of information agrees with our intuition *when* we are in situations typical of computers and the transmission of information as in the telephone, radio, and TV. But it also says that the book (of a given size and type font) with the most information is the book in which the symbols are chosen uniformly at random! Each symbol comes as a complete surprise.

Randomness is *not* what humans mean by information. The difficulty is that the technical definition of the measure of information is connected with *surprise*, and this in no way involves the human attitude that information involves *meaning*. The formal theory of information avoids the "meaning" of the symbols and only considers the "surprise" at the arrival of the symbol, thus you need to be wary of applying information theory to situations until you are *sure* that "meaning" is not involved. Shannon was originally urged to call it *communication theory* but he chose *information theory* instead.

A second fault is that the definition is actually a relative definition— relative to the state of *your* knowledge. For example, if you are looking at a source of random numbers and compute the entropy you get one result, but if I tell you the numbers are from a particular pseudo random number generator then they are easily computed by you and contain no surprise, hence the entropy is 0. Thus the entropy you assign to a source depends on your state of knowledge about it—there is no meaning to the words "the entropy of a source" as such!

Example 7.4–2   The Entropy of a Single Binary Choice

It very often happens that you face a binary choice, yes–no, good–bad, accept–reject, etc. What is the entropy for this common situation? We suppose that the probability of the first one is p, and hence the alternate is $q = 1 - p$. The entropy is then

$$H(p) = p \log \frac{1}{p} + q \log \frac{1}{q} = H(q)$$

where again we use the probability p as the argument of H since there is only one value to be given; for the general distribution we must, of course, give the whole distribution (minus, if you wish, one value due to the sum of all the $p(i)$ being 1), and hence we generally use the name X of the random variable as the argument of $H(X)$.

Example 7.4–3   The Entropy of Repeated Binary Trials

Suppose we have a repeated binary trial with outcome probabilites p and q, and consider strings of n trials. The typical string has the probability

$$ppqqpqp \ldots pqq = p^k q^{n-k}$$

and for each k there are exactly $C(n, k)$ different strings with the same probability. We have, therefore, that the entropy of the distribution of repeated trials is

$$H = \sum C(n, k) p^k q^{n-k} \log \frac{1}{p^k q^{n-k}}$$

We could work this out directly in algebraic detail and get the result, but we know in advance, from the result (7.3–5), that it will be (in the binary choice we can and do use just p as the distribution)

$$H = nH(p) = nH(q) = n \left[p \log \frac{1}{p} + q \log \frac{1}{q} \right] \qquad (7.4\text{–}2)$$

We may define a *conditional entropy* when the events are not independent. Consider two related random variables X and Y. We have, dropping indices since they are obvious,

$$p(x, y) = p(x) p(y \mid x)$$

Thinking of the distribution $p(y \mid x)$ for a fixed x we have

$$H(Y \mid x) = \sum_y p(y \mid x) \log \frac{1}{p(y \mid x)}$$

as the *conditional entropy of y given x*. If we multiply these conditional entropies by their respective probabilities of occurance $p(x)$ we get

$$H(Y\mid X) = \sum p(x)p(y\mid x)\log\frac{1}{p(y\mid x)}$$

$$= \sum p(x,y)\log\frac{1}{p(y\mid x)}$$

Thus entropies may be computed in many ways and are a flexible tool in dealing with probability distributions.

The entropy is the mean of $\{-\log p(i)\}$; what about the variance of the distribution? We have easily

$$\text{Var} = E\left\{\left[\log\frac{1}{p(i)} - H\right]^2\right\} = E\{\log^2 p(i)\} - H^2 \qquad (7.4\text{-}3)$$

Example 7.4–4   *The Entropy of the Time to First Failure*

The time to the first failure is the random variable (p = probability of success)

$$X = \{p^{n-1}q\} \qquad (n = 1, 2, \ldots)$$

If we compute the entropy we find

$$H(X) = \sum_{n=1}^{\infty} p^{n-1}q\log\frac{1}{p^{n-1}q}$$

Shifting the summation index by 1 we get (using equations (4.1–2) and (4.1–4))

$$H(X) = q\sum_{n=0}^{\infty} p^n\left\{\log\frac{1}{q} + n\log\frac{1}{p}\right\}$$

$$= q\log\frac{1}{q}\sum_{n=0}^{\infty} p^n + q\log\frac{1}{p}\sum_{n=0}^{\infty} np^n \qquad (7.4\text{-}4)$$

$$= q\log\frac{1}{q}\left\{\frac{1}{1-p}\right\} + q\log\frac{1}{p}\left\{\frac{p}{(1-p)^2}\right\}$$

$$= \log\frac{1}{q} + \frac{p}{q}\log\frac{1}{p} = \frac{1}{q}H(p)$$

The above results depend on being a discrete countably infinite distribution. But for some discrete *infinite* distributions the entropy does not exist (is infinite). We need to prove this disturbing property. Because of the known theorem that for monotone functions the improper integral and infinite series converge and diverge together, we first examine the improper integrals (using $\log_e = \ln$)

$$\int_2^\infty \frac{1}{x \ln x}\, dx = \ln \ln x \Big|_2^\infty \to \infty$$

$$\int_2^\infty \frac{1}{x \ln^2 x}\, dx = \frac{-1}{\ln x}\Big|_2^\infty = \frac{1}{\ln 2}$$

From these we see that

$$\sum_2^\infty \frac{1}{n \log n} \to \infty$$

$$\sum_2^\infty \frac{1}{n \log^2 n} = A \qquad \text{(a finite number).}$$

$$(7.4\text{--}5)$$

We use these results in the following Example.

Example 7.4–5   *A Discrete Distribution with Infinite Entropy*

The probability distribution X with

$$p(n) = \frac{1}{A(n \log^2 n)} \qquad (n = 2, 3, \ldots)$$

has a total probability of 1 (from the above equation (7.4–5) $A > 0$). But the entropy of X is

$$H(X) = \sum_{n=2}^\infty p(n) \log \frac{1}{p(n)}$$

$$= \sum_{n=2}^\infty p(n) \log\{An \log^2 n\}$$

$$= \frac{1}{A} \sum_2^\infty \frac{1}{n \log^2 n}\{\log A + \log n + 2 \log \log n\}$$

$$= \frac{1}{A} \sum_2^\infty \left\{ \frac{\log A}{n \log^2 n} + \frac{1}{n \log n} + \frac{2 \log \log n}{n \log^2 n} \right\}$$

The first term converges; the second term diverges; and the third term is positive so there is no possibility of cancellation of the infinity from the second term—the entropy is infinite.

Thus there are discrete probability distributions which have no finite entropy. True, the distributions are slowly convergent and the entropy slowly diverges, but it is a problem that casts some doubt on the widespread use of the entropy function without some careful considerations. We could, of course, by fiat, exclude all such distributions, but it would be concealing a basic feature that the entropy of discrete probability distributions with an infinite number of terms can be infinite. Similar results apply to continuous distributions, either for infinite ranges or for a finite range and infinite values in the interval (use for example the simple reciprocal transformation of variables).

Exercises 7.4

7.4–1   Plot $H(p)$.

7.4–2   Complete the detailed derivation of the entropy of the binomial distribution.

7.4–3   What is the entropy of the distribution $p(n) = A'/n^{3/2}$?

7.4–4   Find the entropy of the distribution $p(n) = (1 - a)a^n$, $n = 0, 1, 2, \ldots$

7.5   The Maximum Entropy Principle

The *maximum entropy principle* states that when there is not enough information to determine the whole probability distribution then you should pick the distribution which maximizes the entropy (compatible with the known information, of course).

In a sense the maximum entropy principle is an extension of the principle of indifference, or of consistency (see Section 6.4), since with nothing known both principles assign a uniform distribution (Example 7.4–1). When some facts are known they are first used, and then the principle of maximum entropy is invoked to determine the rest of the details of the distribution. Again (see the next Example), the result is reasonable, and the independence is usually regarded as less restrictive than the assumption of some correlations in the table. Thus the maximum entropy assumes as much independence as can be obtained subject to the given restrictions.

Example 7.5–1   *Marginal Distributions*

Suppose we have a distribution of probabilities $p(i,j)$ depending on two indices of ranges m and n. We are often given (or can measure) only the marginal distributions, which are simply the row sums $R(i)$

$$\sum_j p(i,j) = R(i) \qquad \text{with} \qquad \sum_i R(i) = 1$$

and the column sums $C(j)$

$$\sum_i p(i,j)] = C(j) \qquad \text{with} \qquad \sum_j C(j) = 1$$

where, of course, the indicated arguments i and j may have different ranges. Thus we have $(m+n-2)$ independent numbers (satisfying the restrictions on the row and column sums), and wish to assign reasonable values to the whole array $p(i,j)$ consisting of $mn-1$ values, (restricted by $\Sigma_{i,j} p(i,j) = 1$). There is, of course, no unique answer to such situations for $m, n > 1$.

The maximum entropy rule asserts that it is reasonable to assign the values to the $p(i,j)$, subject to the known restraints, so that the total entropy is maximum. We begin with Gibbs' inequality

$$\sum_i^j p(i,j) \log \frac{R(i)C(j)}{p(i,j)} \leq 0$$

By the Gibbs' inequality the entropy is a maximum when all the

$$p(i,j) = R(i)C(j) \qquad\qquad (7.5–1)$$

This is the same thing as assuming that the probabilities in i and j are independent—the least assumption in a certain sense!

We now look as a classic example due to Jaynes [J]. It has caused great controversy mainly because people cannot read what is being stated but only what they want to read, (and Jaynes tends to be belligerent).

Example 7.5–2   *Jaynes' Example*

Suppose the sample space has six points with values 1, 2, ..., 6. We are given the information that the average (expected value) is not 7/2 (as it would be if the probabilities were all 1/6) but rather it is 9/2. Thus we have the condition

$$E\{X\} = \sum_{i=1}^{6} ip(i) = 9/2 \tag{7.5-2}$$

The maximum entropy principle states that we are to maximize the entropy of the distribution

$$H\{X\} = \sum_{i=1}^{6} p(i) \log \frac{1}{p(i)}$$

subject to the two restrictions: (1) that the sum of the probabilities is exactly 1,

$$\sum_{i=1}^{6} p(i) = 1 \tag{7.5-3}$$

and (2) that (7.5–2) holds.

We use Lagrange multipliers and set up the function

$$L\{p(i)\} = H\{X\} - \lambda \sum_{i=1}^{6} \{p(i) - 1\} - \mu \sum_{i=1}^{6} \{ip(i) - 9/2\} \tag{7.5-4}$$

We next find the partial derivatives of $L\{p(i)\}$ with respect to each unknown $p(i)$, and set each equal to zero,

$$-1 - \log p(i) - \lambda - \mu i = 0 \quad (i = 1, \ldots, 6) \tag{7.5-5}$$

This set of equations (7.5–5) together with the two side conditions (7.5–2) and (7.5–3) determine the result. To solve them we first rewrite (7.5–5)

$$\log p(i) = -(1 + \lambda + \mu i)$$
$$p(i) = e^{-(1+\lambda+\mu i)} \tag{7.5-6}$$

To simplify notation we set

$$e^{(1+\lambda)} = A, \quad e^{-\mu} = x, \tag{7.5-7}$$

In this notation the three equations (7.5-6), (7.5-3) and (7.5-2) become

$$p(i) = \frac{1}{A} x^i \tag{7.5-8}$$

$$\sum_{i=1}^{6} x^i = A \tag{7.5-9}$$

$$\sum_{i=1}^{6} i x^i = \frac{9}{2} A \tag{7.5-10}$$

To eliminate A divide equation (7.5-10) by equation (7.5-9)

$$\frac{9}{2} = \frac{\displaystyle\sum_{i=1}^{6} i x^i}{\displaystyle\sum_{i=1}^{6} x^i}$$

The solution $x = 0$ is impossible, and we easily get

$$\sum_{i=1}^{6} \left(\frac{9}{2} - i \right) x^{i-1} = 0 \tag{7.5-11}$$

Expanding the summation and clearing of fractions we get the following algebraic equation of degree 5 in x,

$$3x^5 + x^4 - x^3 - 3x^2 - 5x - 7 = 0 \tag{7.5-12}$$

There is one change in sign and hence one positive root which is easily found by the bisection method (or any other root finding method you like). From equation (7.5-9) we get A, and from equation (7.5-8) we get the geometric progression of probabilities $p(i)$ as follows.

$$p(1) = 0.05435$$
$$p(2) = 0.07877$$
$$p(3) = 0.11416$$
$$p(4) = 0.16545$$
$$p(5) = 0.23977$$
$$p(6) = 0.34749$$

Is the result reasonable? The uniform distribution $p(k) = 1/6 = 0.16666$ has a mean $7/2$ that is too low; this distribution has the right mean of $9/2$.

To partially check the result we alter the 9/2 to 7/2 where we know the solution is the uniform distribution. The equation (7.5–11) becomes

$$\sum_{i=1}^{6} \left(\frac{7}{2} - i \right) x^{i-1} = 0$$

and this upon expanding and multiplying by 2 is (compare with equation (7.5–12))

$$5x^5 + 3x^4 + x^3 - x^2 - 3x - 5 = 0$$

which clearly has the single positive root $x = 1$. From equation (7.5–9) $A = 6$ and from equation (7.5–8) $p(i) = 1/6$ as we expect.

When we examine the process we see that the definition for A arose as "the normalizing factor" to get from the x^i values to the probability values $p(i)$, and it arose from the combination of the entropy function definition plus the condition that the sum of the probabilities is 1. Since these occur in most problems that use the maximum entropy criterion for assigning the probability distribution, we can expect that it will be a standard feature of the method of solution. The root x, see equation (7.5–7), arose from the condition on the mean; other side conditions on the distribution will give rise to other corresponding simplifying definitions. But so long as the side conditions are linear in the probabilities (as are the moments for example), we can expect similar methods to work.

If you do not like the geometric progression of probabilities that is the answer (from the use of the maximum entropy principle) then with nothing else given what other distribution would you like to choose? And how would you seriously defend it?

Example 7.5–3 *Extensions of Jaynes' Example*

In order to understand results we have just found, it wise to generalize, to extend, and otherwise explore the Example 7.5–2. We first try not 6 but N states, and assign the mean M. Corresponding to (7.5–2) we have

$$\sum_{i=1}^{N} ip(i) = M$$

and in (7.5–4) we have, obviously M in place of 9/2. The set of equations (7.5–5) are the same except there are now n of them. The change in notation equation (7.5–7) is the same. The corresponding three equations are now

$$\sum_{i=1}^{N} x^i = A$$

$$\sum_{i=1}^{N} ix^i = MA$$

$$p(i) = \frac{1}{A}x^i$$

The elimination of A gives corresponding to equation (7.5–11)

$$\sum_{i=1}^{N}(M - i)x^{i-1} = 0$$

Again there is one positive root. (The root $= 1$ only if M is $(1 + 2 + \cdots + N)/N = (N + 1)/2$. The rest is the same as before.

Another extension worth looking at is the inclusion of another condition on the sum (the second moment)

$$\sum_{i=1}^{N} i^2 p(i) = S$$

We will get the Lagrange equation

$$L\{p(i)\} = H(S) - \lambda \sum_{i=1}^{N}\{p(i) - 1\} - \mu \sum_{i=1}^{N}\{ip(i) - M\} - \nu \sum_{i=1}^{N}\{i^2 p(i) - S\}$$

The resulting equations are

$$\log p(i) = -(1 + \lambda + \mu i + \nu i^2)$$

The same substitutions (7.5–7) together with $e^{-\nu} = y$ give us

$$\sum_i x^i y^{i^2} = A$$

$$\sum_i i x^i y^{i^2} = MA$$

$$\sum_i i^2 x^i y^{i^2} = SA$$

We eliminate A by dividing the second and third equations by the first equation, and we have a pair of equations

$$\sum_i (M - i) x^i y^{i^2} = 0$$

$$\sum_i (S - i^2) x^i y^{i^2} = 0$$

To solve this pair of nonlinear equations we resort to a device [H, p.97] that is useful. Call the first expression $F(x, y)$ and the second $G(x, y)$. We want a common solution of $F = 0$ and $G = 0$. Now for each pair of numbers $\{x, y\}$ there is a pair of numbers $\{F(x, y), G(x, y)\}$ and this point falls in some quadrant. Plot in the x–y plane the corresponding quadrant numbers of the point (F, G). The solution we want is clearly where in the F, G variables the four quadrants meet. Thus a crude plot of the quadrant values will indicate this point. A refined plot at a closer spacing in the x–y variables will give a more accurate estimate of the solution. Successive refinements will lead you to as an accurate solution as you can compute. It is not necessary to plot a large region of the x–y plane, you only need to track one of the edges dividing the quadrants, either $F = 0$ or $G = 0$, until you meet the other edge dividing the quadrants.

7.6 Summary

In all probability problems there is always the fundamental problem of the initial probability assignment. In the earlier chapters we generally assumed, from symmetry, that the distribution was uniform, but there are times when we need to do something else since the uniform distribuition may not be compatible with the given facts—such as in the above Example 7.5–2 where the mean is 9/2. The maximum entropy principle allows us to make an assignment of the probabilities that *in a reasonable sense* makes the least additional assumptions about the interrelationships of the probabilities.

That the principle is sometimes useful does not make it the correct one to use in all cases. Nor should the name "entropy" imply any overtones of physical reality. You need to review in your own mind how we got here, what the principle seems to imply, and consider what other principles you might try. That in maximum entropy problems the initial distribution is not completely given is actually no different than when we invoked the principle of indifference (the principle of consistency) to get from no probabilities assigned to the uniform distribution. In those cases where not enough information is given to determine all the details of the initial distribution then something must be assumed to get it. To assert that nothing can be done is not likely to be satisfactory. Nor is the statement that you can get almost anything you want. Sensible action is required most times.

To help you resolve your problem of accepting or rejecting this maximum entropy principle (which is compatible with the principles we have so far used) we have included a great deal of background information, including derivations of possibly unfamiliar results. We have also included some of the details we used in the Lagrange multiplier approach. It is a familiar method in principle, though in practice it often involves a great deal of hard arithmetic and algebra. But can you really hope to escape such details in complex problems? The labor of computation is not a theoretical objection to the principle—but it may be an objection in practice! How badly do you want a reasonable basis for action? After all, we now have powerful computing machines readily available so it is programming effort rather than computing effort that is involved.

You need to adopt an open mind towards this maximum entropy method of assigning the initial distribution. But an open mind does not mean an empty head! You need to think carefully before either applying or abandoning the maximum entropy principle in any particular case. For further material see [L–T].

8

Models of Probability

"All possible "definitions" of probability fall far short of the actual practice."   Feller [F, p. 19]

8.1   General Remarks

Most authors of probability books present one version of probability as if it were the true and only one. On the contrary, we have carefully developed a sequence of models based on what seems to be a reasonably scientific approach. However, there are other significantly different models that need to be thought about, and the main purpose of this chapter is to present some of them. It should not surprise the reader that occasionally in the middle of my presentation of a model I defend the ones used in this book. Part of the explanation is that various models have different underlying motivations (often statistical), some of which are quite different from the ones this book uses; we assume that probability should provide a basis for reasonable actions in the real world and not just be admired. See [Ke] for a careful discussion of many models of probability.

> *"We need flexibility to understand the world around us, and we gain such flexibility by having flexible viewpoints stressing different aspects. This flexibility is not easily reconciled with a rigid axiomatic framework, such as may be found in a traditional course on rational mechanics."*   Gregory Wannier [Wa, p.1–2]

We have stressed that the assignment of the initial probabilities is of fundamental importance in the applications of probability. Most texts blithely

[279]

assume that this is not part of their problem; rather like most mathematical texts, where problems are posed in a formal mathematical setting, probabilities are either actually given or strongly indicated. Rarely do even the "word problems" require much translation from the natural language into mathematical symbols—and they are usually highly artificial anyway. We are concerned with the *use* of probability. Because the background knowledge of most applications is fundamental to practical work, we are unable to include such material in this book.

Very early we noted the importance of decisions based on probability and developed an approach that seems to bear some relation to actions taken in the real world. Unfortunately, many applications are founded on a far less firm foundation, yet the decisions taken in government, law, medicine, etc. constantly affect our lives for better or worse. It is therefore necessary to display clearly some of their assumptions—they are often not the same as those we have adopted and checked with various gambling and related situations; indeed many have significantly subjective elements in them. Science tries to be objective, though it cannot succeed in this as much as most people wish. Still, it is a goal that in the past has seemed highly desirable and highly productive. Perhaps in the future science may embrace a much more subjective approach—but to do so would mean the abandonment of the standard of reproducibility by others! Science has not always had perfect reproducibility—for example, in observational astronomy the locations of the planets are not exactly reproducible since the observatories are not in the same places on earth nor are the planets at various times in the sky (they are not perfectly periodic). Still, astronomy, which is the earliest of the sciences, managed to develop under the twin handicaps of no experimental control and no exact reproducibility.

We have also stressed robustness of our models. If the models is not robust then it is doubtful that it can be useful. It is a topic that by tradition has been greatly neglected in discussions of probability.

8.2 Review

We began with very simple situations, the toss of a coin, the roll of a die, or the draw of a card—all typical gambling situations where there is an underlying symmetry and the equipment has been carefully designed (for independence and a uniform distribution) and well tested by long experience. We assumed exact symmetry and argued that unless we assigned the same measure to each of the interchangable events then we would be embarrassed by inconsistency. Thus we were forced, by the apparent symmetry, to assign the same probability to each interchangable event. We also made the total probability exactly equal to 1.

From this simple model we derived the "Weak law of large numbers" which states that the average of a large number of independent binary trials is close to the original assigned probability—probably! But this assumed the existence of the mean of the distribution (see Appendix 2.A). In Section 8.7 we will look at this point more closely.

This simple model can give only rational numbers for probabilites. If we think of tossing a thumb tack to see if it lands point up or not, (Figure 1.10–1), then we are not prepared to compute any underlying symmetry and are forced to estimate this probability, which we believe exists independent of the actual trials—though perhaps you should consider this point carefully before going on, (see Section 8.9). This probability involves not only the thumb tack, but also the surface on which it falls as well as the complex tossing mechanism. We must even consider the angular momenta about the various axes. For example, suppose you had a long rod with n symmetrically placed flat (planar) sides each of the same size. In tossing the rod you would not expect it to come to rest on either end of the rod, but rather you would assign equal probability (due to the supposed symmetry) to each side face. But now think of the rod being shortened until it is a thin, flat disc—the shape of a coin. Now you would reject the toss ending on a side and you would assign equal probability to each face. When the rod is long a large angular momentum about the long axis forces the rod come to rest on one of the long sides and forbids the ends as final positions. With the coin it is the opposite. Hence there are lengths of the rod where the angular momenta affect the results significantly. In a randomization process the entire mechanism that makes the random selection requires close attention—which it gets in gambling places! In practice people are often careless about this point; it is usually hard to get a genuine random sample!

In passing from games to production lines, we may not know all the details of the process but we have enough familiarity with such mechanisms to feel that the local interchangability of the parts coming off the production line is often a reasonable assumption—but that there are also effects that change slowly with time, such as the wear and tear of parts of the machines, the slow changing of adjustments, etc. Indeed, it is exactly this which is the basis of *Quality Control*. We can expect small fluctuations from item to item, but any trends in the output (revealed by falling outside the *control limits*) should be examined and corrected before the product becomes unacceptable.

However, the items going through a production line and the patients going through a hospital, for example, are rather different. We have less belief in the interchangeability of the patients, and of the constancy of the hospital processes. The statisticians have a word for it, "stationarity," which assumes that there are no significant changes in the statistics as a function of time. Now we face a couple of troubles—there is no symmetry to imply an underlying probability, and we can only wonder about the changing effects in time. For a hospital, and such situations, the quality control approach is hard to apply.

Thus we see that the well-founded, well-tested, simple model of probability based on gambling situations loses a lot of its reliability when we pass to other situations. Yet we continue to use probability arguments in medical and other social situations. We are forced to do so since we have nothing else, and to do nothing in the face of uncertainty is as positive an action as any other particular choice. Hence we must look at other probability models; the two quotations in Section 8.1 apply. Whether we like it or not, many different probability models must be examined at least briefly, and this is the purpose of the rest of this Chapter.

Finally, more than a few philosophers have stated that if a probability is less than some preassigned number (Buffon used 1/10,000) then the probability is 0. Whenever very high reliability of a large system is needed then the probability of failure of the individual components must be very small. Moreover, a low probability does not scale in time. Let the expected number of failures be one per day. Then the expected number of failures per microsecond is

$$1/\{24 \times 60 \times 60 \times 10^6\} = 1/8.64 \times 10^{10} = 1.157 \ldots 10^{-11}$$

Clearly, for events in time the probability must depend on the interval of time used, and small probabilites over long times cannot be ignored. Given enough time anything that is possible will (probably) happen.

8.3 Maximum Likelihood

"There comes a time in the life of a scientist when he must convince himself either that his subject is so robust from a statistical point of view that the finer points of statistical inference he adopts are irrelevant, or that the precise mode of inference he adopts is satisfactory."   A. W. F. Edwards, [Ed, p. xi]

A method of assigning initial probabilities, similar to but different from maximum entropy, is that of *maximum likelihood*, which we will illustrate by some examples. We have used the maximum entropy criterion to determine a whole probability distribution; we will use the maximum likelihood criterion to determine a parameter in a distribution. It also resembles the weak law of large numbers in the sense that it assigns a probability based on past data.

Example 8.3–1   *Maximum Likelihood in a Binary Choice*

Given a simple binary choice (a Bernoulli trial) which you observe had k successes out of n trials ($k \leq n$), what is a reasonable assignment for the probability p?

Evidently p can be any number between 0 and 1, and we must make a choice. Bernoulli himself suggested, based on the formula for k successes in n independent trials each with probability p, see equation (2.4–2),

$$b(k; n, p) = C(n, k)p^k q^{n-k}$$

that the value of p chosen should be the one that makes this probability a maximum, *the maximum likelihood* choice of p. How could one seriously propose to adopt a value of p which makes the actual observations highly unlikely as compared to some other value of p?

To find the maximum we differentiate with respect to the variable p, set the derivative equal to zero, and solve for the value of p. We can neglect, in this process, any front coefficients. We get for the derivative

$$kp^{k-1}q^{n-k} + (n-k)p^k(1-p)^{n-k-1}(-1) = 0$$

or (if $pq \neq 0$)

$$k(1-p) - (n-k)p = 0$$

Expand and cancel the like terms

$$k - kp - np + kp = k - np = 0$$

From this we get

$$p = k/n$$

This is the ratio of the successes to the total number of trials and is exactly what the normal person using the weak law of large numbers would suggest! In this simple example the maximum likelihood method agrees with common sense.

Notice that this result did not use the weak law of large numbers; it is a result of the method of maximum likelihood. In this case the two methods of assigning probabilities agree.

Example 8.3–2   *Least Squares*

Suppose we have made n independent selections $(i = 1, 2, \ldots, n)$ from a Gaussian distribution

$$p(x) = \frac{1}{\sigma\sqrt{2\pi}} e^{-(x-x_o)^2/2\sigma^2}$$

where we assume that the mean x_o is not known *but that the variance is known*. The likelihood of seeing this result for the n independent trials is the product of the probabilities of the individual observations, namely

$$L(x_1, x_2, \ldots, x_n) = p(x_1)p(x_2)\ldots p(x_n)$$

$$= \frac{1}{(2\pi)^{n/2}} e^{-\frac{1}{2\sigma}\sum_{i=1}^{n}(x_i-x_0)^2}$$

According to the maximum likelihood principle we want to maximixe this, namely make the term in the exponent

$$\sum_{i=1}^{n}(x_i - x_0)^2$$

a minimum—and this is exactly the principle of least squares!

Thus the assumption of a Gaussian distribution with a given variance, along with the independence of the events and the principle of maximum likelihood, yields the well respected *principle of least squares*.

Here we have a principle for picking a parameter value in a distribution. The principle tacitly assumes that there is a single peak (unimodal) in the probability density function, though of course it can mechanically be applied to any distribution. In Example 8.3–1 the principle assigned a single probability and in Example 8.3–2 assigned a parameter of a distribution. The maximum likelihood estimate is insensitive to any outliers of a distribution (unlike the mean which is!) and as such is a robust estimator.

Maximum likelihood differs from the earlier maximum entropy. Both criteria, in the places where we have used them—maximum likelihood for whole distributions and maximum likelihood for parameter values of a distribution—have given reasonable results. Whether or not they generally give the same results would need much more exploration than we can give here.

Example 8.3–3   *Scale Free*

If we make a strictly monotone transformation of the independent variable from x to y, we do not affect the maximum likelihood estimate.

This simple fact follows from the standard formula

$$\frac{dF}{dx} = \frac{dF}{dy}\frac{dy}{dx}$$

When $dF/dx = 0$, since by hypothesis (strictly monotone) dy/dx cannot be zero, it follows that dF/dy must be zero.

This is important. Recall that in Chapter 6 (Uniform Probability Assignments) we worried about whether to assign the uniform distribution to the variable x, x^2 or $x^{1/2}$, for example. We gave a useful partial answer which applied to random variables that were generated by a modular reduction, but it is not universally applicable. The use of maximum likelihood removes this problem! The independence of scale is one of the attractive features of the maximum likelihood method.

8.4   von Mises' Probability

Richard von Mises proposed a model for probability whose main aim was to provide a sound statistical basis for probability theory. "Thus, a probability theory which does not introduce from the very beginning a connection between probability and relative frequency is not able to contribute anything to the study of reality." [vM p. 63]. This is the opposite of what we have done; we defined probability using symmetry and then used it, via the weak law of large numbers as a basis, such as it is, to get the usual probabilities used in statistics.

"I have merely introduced a *new name*, that of *collective*, for sequences satisfying the criterion of randomness." [vM, p. 93]. Thus to be a *collective* means that the statistics of any proper subsequence of the original sequence must have in the limit the statistics you would reasonably expect from a random sequence. Proper subsequences include any that are *prespecified*, such as "take only those after the occurence of three heads in a row," but excludes those like "take only those outcomes which are heads," results which depend on the actual outcomes of the events selected. This is a simple realization of the meaning of the words "independent trials."

His justification for this assumption is that gambling systems have been shown over many, many years to be ineffective, and hence, he says, the *collective* represents reality.

My objections to this approach are many. First, while it is commendable to use reality to form the postulates, it also apparently limits the applications to gambling-like situations. For the applications to other areas how would we demonstrate this remarkable property that the sequence of events forms a collective? The apparently equivalent assumptions of the independence of the trials and the constancy of the probability seem more reasonable.

Second, as any theoretician will admit, when tossing a coin a run of 1000 heads in a row will occur sooner or later; indeed in any infinite sequence of trials a run of 1000 or more heads will occur infinitely often! Hence a collective representing coin tosses (zeros and ones) remains a collective if I put 1000 heads (ones) before the original collective. Theoretically in the long run this front end will be swamped and will not affect the assumed statistics of the subsequences—but the long run is painfully long! Most people would immediately reject such a sequence as not being random (a collective). What is wanted in practice is a pasturized, homogenized, certified random sequence which does not start out with an unlikely fluctuation, but rather allows big fluctuations, if at all, very late in the sequence—just enough to get the effects of randomness. The pure mathematical criterion of random does not meet practical needs. Thus *in general* the long run ratio of successes/total trials approaching the probability is dubious in practice (where we are restricted to comparatively few trials).

If, as in the previous paragraph, we faced a long run of, say, 1000 heads, we would probably soon adopt the hypothesis that $\Pr\{\text{head}\} = 1$, and when we later came to the rest of the data we would probably revise our hypothesis for the value of the probability. Thus in practice we generally have in the back of our minds a dynamic theory of probability (see Section 8.10 on Bayesian statistics). As such it falls outside of this book, and belongs properly to statistics, but the two topics are so closely related that each cannot completely avoid the other without serious loss to itself.

My objection to postulating things that you cannot hope to measure in practice is a broad objection, and includes most theorems whose hypotheses cannot be reasonably known or verified by you for your application; it may be nice mathematics but it is of little use in practice.

This does not mean that the book he wrote [vM] is not worth reading. He was a very intelligent person and had thought long and hard about the problem of justifying statistics; more so than most statisticians! While this present book is not on statistics, every one realizes that any statistics beyond the simple organizing and displaying of data, must rest on some model of probability.

8.5   The Mathematical Approach

"Idealists believe that the postulates determine the subject;
realists believe that the subject determines the postulates."

In 1933 Kolmogoroff laid down some axioms for probability. Let E be a set of elements which are the elementary events, and F is a collection of subsets of E; the elements of the set F are the correponding events. He assumed

    1. F is a field (operations of addition and product).
    2. F contains the set E
    3. To every set A in F there is an nonnegative real number $P(A)$. This number $P(A)$ is called the probability of the event A.
    4. $P(E) = 1$
    5. When A and B are disjoint, then

$$P(A + B) = P(A) + P(B)$$

For infinite probability fields he further assumes that infinite sums and products of sets are sets.

    6. For every sequence

$$A_1 \supset A_2 \supset \cdots \supset A_n \supset \cdots$$

whose intersection is 0 gives

$$\lim_{n \to \infty} P(A_n) = 0$$

All of this amounts to assuming that the sets form a σ-Algebra or a Borel field.

    As one examines these axioms one wonders how, in a particular application, one would verify whether or not they included the right things and excluded the wrong things. Deductions from aspects that are not in the reality being modelled might distort the conclusions. In all applications of probability there are always the two dangers: (1) of not including enough in the model, and (2) of including too much.

    The Kolmogoroff approach is favored by mathematicians and leads immediately to measure theory. It gives an elegant theory with many interesting mathematical results, but one can only wonder to what extent the model represents reality. I have long said, "If whether an airplane will fly or not depends on some function that arose in the design being Lebesgue integrable but not Riemann integrable, then I would not fly in that plane." Dare one act in this real world on such a model, and if so, when and under what circumstances would one be willing to risk one's life on the results from the model?

    If we assume that reality is what you can hope to describe, and what you can never hope to describe is not reality (as is customary, but not always true, in science) then with respect to the real number system measure theory ignores all of reality and contains only unreality. One can only wonder what can be expected from using it when applied to reality.

These are not facetious remarks—they raise a basic question of whether or not probability theory is to be regarded solely as a mathematical art form or mainly as a useful tool in dealing with the real world. Two eminent probabilists, Kac [K,p. 24] and Feller, have raised strong objections to the viewpoint that probability is a branch of measure theory, "Too many important details are lost." is the objection to the sweeping generality that overlooks the important special features of probability problems.

Some of the most quoted examples of the use of Lebesgue integration in probability theory seem on the face of them to be strictly non-applicable since they deny aspects of reality that we believe are right. Trajectories which have no direction at any place seem not to be realistic modelling of Brownian motion, for example, though *perhaps* the idealization is safe to use in many situations—how would you decide?

There is a small heretical (?) group of probabilists [Ra] who do not believe in the infinite additivity assumed in the mathematical approach of Kolmogorov. Their argument is that we have experience with the finite additivity of probability, and hence they believe in modelling only what we have some experience with. No one can possibly have direct experience with the assumed infinite additivity. As a result finite additivity wipes out, among other things, Lebesgue integration and this annoys professional probabilists who worry more about finding new theorems than the relevance to reality. Thus some purists assert, "Finite additivity is not part of probability theory."

What are the dangers of assuming infinite additivity? Simply that you may be including effects that are not mirrored, even slightly, in reality. My favorite example of this sort of thing is the well known theorem, widely cited by mathematicians, that with ruler and compass (in classical Euclidean geometry) you cannot trisect an arbitrary angle. The fact is that with two marks on the ruler you can. Thus the truth or falsity of this theorem rests on a very slight difference (indeed in a practical sense almost trivial difference) in the idealizations being made by the corresponding postulates. So too, for those who believe only in the finite additivity there will be theorems proved in the infinite additivity model that are artifacts of this detail and no amount of elegant mathematics can compensate for the induced lack of the reality in what follows, so they say.

It was just such thinking that now drives me, as cited in the opening quotation of this Chapter, to a reconsideration of the probability models I have used in the past (sometimes without having really carefully thought before acting). It now seems to me that no single model of probability would have been appropriate to all the different situations I faced, and that probably no one model will do for the future either. Apparently it is necessary in every application to think through the relevance of the chosen probability model. Even the use of the real number system, with all its richness and known peculiar features, needs to be questioned regularly (see also Sections 5.1 and 8.13).

For "non-Kolmogorovian" probability models, and the role of Bayes' theorem, see for example [D. p. 297–330]

8.6 The Statistical Approach

In probability theory, strictly speaking, we go from initial probabilities and probability distributions to other derived ones. In statistics we often go from the observed distributions to other inferred prior distributions and probabilities. The model of probability you adopt is often relevant to what actions you later take, though most statistics books do not make plain and clear just what is being assumed. Unfortunately, many times the statistical decisions affect our lives, from health, medicine, pollution, etc. to reliability of power plants, safety belts vs. air bags, space flights, etc.—too numerous to mention in detail, and in any case the list is constantly growing.

The weak law of large numbers encourages one to think that the average of many repetitions will give an estimate of the underlying probability. But we have seen (as in Example 2.9–2) that the number of trials necessary to give an accurate estimate of the probability may be much more than we can ever hope to carry out. And there are further difficulties. We assumed the existence of the mean; suppose it does not exist! (see the next Section 8.7).

It is not that the author is opposed to statistics—there is often nothing else to try in a given situation—but the stated reliabilities that statisticians announce are often not realized in practice. Unfortunately, the foundations of statistics, and its applicability in many of the situations in which it is used, leaves much to be desired, especially as statistics will probably play an increasingly important role in our society.

8.7 When the Mean Does Not Exist

When the mean does not exist then we cannot depend on the weak law of large numbers (see Appendix 2.A), and we clearly cannot depend on the usual statistical process of averaging n samples to get an good guess for the mean of the distribution. Indeed, there is a well known counterexample.

Example 8.7–1   *The Cauchy Distribution*

The Cauchy distribution

$$p(x) = \frac{1}{\pi(1+x^2)} \tag{8.7–1}$$

has the property that the distribution of the average of n samples has the same, original distribution.

To show this by elementary (though tedious) calculus we begin with the sum of two similar variables from Cauchy distributions, one from the Cauchy distribution with parameter a

$$p_1(x) = \frac{a}{\pi(x^2+a^2)}$$

and the second from the distribution with parameter b

$$p_2(x) = \frac{b}{\pi(x^2+b^2)}$$

The distribution of the sum of the two samples, one from each distribution, is given by the convolution integral

$$p(x) = \int_{-\infty}^{\infty} p_1(s)p_2(x-s)\,ds$$

$$= \frac{ab}{\pi^2} \int_{-\infty}^{\infty} \frac{1}{(s^2+a^2)[(x-s)^2+b^2]}\,ds \tag{8.7–2}$$

The reader who wants to skip the mathematical details of the integration may go directly to equation (8.7–9).

To integate this we apply the standard method of partial fractions. From a study of the third degree terms in s, and the later integration steps, we are led to assume the convenient partial fraction form

$$\frac{1}{(s^2+a^2)[(x-s)^2+b^2]} = \frac{As+B}{s^2+a^2}$$

$$+ \frac{A(x-s)+C}{(x-s)^2+b^2} \tag{8.7–3}$$

Suppose for the moment that we have found the A, B, and C. We then do the integrations

$$\frac{ab}{\pi^2} \int_{-\infty}^{\infty} \left\{ \frac{As}{s^2 + a^2} + \frac{B}{s^2 + a^2} + \frac{A(x - s)}{(x - s)^2 + b^2} + \frac{C}{(x - s)^2 + b^2} \right\} ds$$

$$= \frac{ab}{\pi^2} \left\{ \frac{A}{2} \ln \left[\frac{s^2 + a^2}{(x - s)^2 + b^2} \right] + \frac{B}{a} \arctan \frac{s}{a} - \frac{C}{b} \arctan \frac{x - s}{b} \right\} \Big|_{-\infty}^{\infty}$$

The *ln* term drops out since at both ends of the range the argument is 1 and the function is continuous in the whole range. The first Arctan gives π and the second the same result only with a minus sign. Hence we have for the convolution $p(x)$, equation (8.7–2),

$$p(x) = \frac{bB + aC}{\pi} \tag{8.7-4}$$

We now carry out the determination of the coefficients of the partial fraction expansion (8.7–3). Clearing equation (8.7–3) of fractions we have the identity in s

$$1 = (As + B)\{(x - s)^2 + b^2\} + (Ax - As + C)\{s^2 + a^2\} \tag{8.7-5}$$

The cubic terms in s cancel (as they should due to the form we assumed). The quadratic terms in s lead to

$$0 = -2Ax + B + Ax + C \Rightarrow Ax - B - C = 0 \tag{8.7-6}$$

Using $s = 0$ in equation (8.7–5) we get

$$1 = xa^2 A + (x^2 + b^2)B + a^2 C \tag{8.7-7}$$

Using $s = x$ in equation (8.7–5) we get

$$1 = xb^2 A + b^2 B + (x^2 + a^2)C \tag{8.7-8}$$

Eliminate A (from equation (8.7–4) we do not need A) using equation (8.7–6), and set

$$x^2 + a^2 + b^2 = K.$$

We get for the two equations (8.7–7) and (8.7–8)

$$KB + 2a^2 C = 1$$

$$2b^2 B + KC = 1$$

The determinant of the system of equations

$$K^2 - 4a^2b^2 = x^2 + (a - b)^2$$

is not zero, and we get for equation (8.7–4)

$$\frac{bB + aC}{\pi} = \frac{b(K - 2a^2) + a(K - 2b^2)}{\pi(K^2 - 4a^2b^2)}$$

$$= \frac{(a + b)(K - 2ab)}{\pi(K + 2ab)(K - 2ab)}$$

$$= \frac{a + b}{\pi\{x^2 + (a + b)^2\}}$$

Thus we have, finally, for equation (8.7–2)

$$p(x) = \frac{a + b}{\pi\{x^2 + (a + b)^2\}} \tag{8.7–9}$$

This shows that the two parameters merely add in the convolution operation.

With (8.7–9) as a basis for induction we conclude that the sum of n samples from the original Cauchy distribution equation (8.7–1) will have the parameter n. We want the average so we need to replace the variable x by x/n—and we get the original distribution back! Thus sampling from the Cauchy distribution and averaging gets you nowhere—one sample has the same distribution as the average of 1000 samples!

The reason for this is that the tails of the Cauchy distribution are large—and taking another sample means, among other things, that the new sample is too likely to fall far out and distort the computed mean of the samples.

If we take an even larger tailed distribution we get even worse results. Suppose we take a distribution

$$p(x) = \frac{C(p)}{1 + |x|^p} \qquad (1 < p < 2), \qquad (-\infty, \infty) \tag{8.7–10}$$

where $C(p)$ is a suitable constant, which we now find, that makes the total probability $= 1$. By symmetry the integral of $p(x)$ is

$$2C(p) \int_0^\infty \frac{1}{1 + x^p}\, dx = 1$$

Set $x^p = t$, that is $x = t^{1/p}$. We get

$$\frac{2C(p)}{p} \int \frac{t^{1/p-1}}{1 + t}\, dt = \frac{2C(p)}{p} \frac{\pi}{\sin \frac{\pi}{p}} = 1$$

from a well known integral. Hence we have the $C(p)$ of the distribution. But we see, at least intuitively, that for $(1 < p < 2)$ the sampling and averaging process will give results that are worse (in the sense of computing an average as indicated by the weak law of large numbers) than taking one sample and using it alone!

> *"In the modern theory variables without expectation play an impor-*
> *tant role and many waiting and recurrence times in physics turn*
> *out to be of this type. This is true even of the simple coin-tossing*
> *game."*   Feller [F, vol 1, p. 246]

Evolution seems to have given us the ability to recognize patterns with repetition when the weak law of large numbers applies, but when the source distribution of events has large enough tails we simply cannot see the pattern. This, in turn, suggests that in many social situations, say our own lives, we do not see the underlying patterns because the distribution of the apparently random events is too variable. Hence we need to think a little about what to do in similar situations. If we transform the independent variable (the way we measure the effect) by a *nonlinear* contraction then we may have an expected value in that variable and hence a weak law of large numbers—we must reduce the variability of the observations if we are to find patterns. For example, in equation (8.7–10) with $p = 3/2$ we can use

$$x = t^2$$

and work with the distribution density

$$\frac{C(3)}{1 + |t|^3}$$

When we use this probability density function the mean exists and we can invoke the weak law of large numbers. The actual contracting function to use depends on the distribution—too strong a contraction will remove all the details, and too weak a one will barely bring in the weak law of large numbers and the basis of statistics.

Exercises 8.7

8.7–1 Show that for the Cauchy distribution the probability of falling in the interval $-1 \leq x \leq 1$ is $1/2$

8.8 Probability as an Extension of Logic

In logic statements are "yes-no," and classical logic has no room for "degrees of truth." Many people, beginning in the earliest years of probability theory, have tried using probability to extend logic to cover *the degree of truth*. Such problems as the lying of witnesses in court, and the number of persons to serve on a jury, for example, have been studied under various hypotheses as to reliability of the individuals.

It is evident that we do have degrees of belief, but it is less evident that they can be successfully combined into a formal system for manipulating the probabilities, let alone that the original quantification of the intuitive beliefs can be made reliable. It is not that the same rules of deduction that we use in gambling situations can not be used, what is doubted is the degree to which you can expect to act successfully on such computations.

When you consider how difficult it is to get probability values from observed data, as the statistician is forced many times to do, one wonders just what faith people could have placed in social applications such as estimating the probability of lying or the size to use for a jury. What they did was to *assume* a certain model, and then make rigorous mathematical deductions from these assumptions without regard to the robustness of the model. For example, if a witness was found to have lied 9 times out of ten in the past (and how would anyone ever verify this in practice?) then they would assign a probability of 1/10 for the truth of the next statement made by the witness. Most sensible people would ask about the self-interest of the witness in each statement before ever making any such judgement on a particular statement made. Similarly with juries. The idea that there is a total independence of the individuals on a jury as they make up their minds is foolish, but rather there is, from jury to jury depending on local circumstances, some correlations between some members. Too much depends on the particular circumstances, the actual jury members, the specific case, and the personalities of the lawyers and witnesses, to try to make any serious statements beyond wild guesses. The robustness of the model is simply being ignored.

But many well respected authors have tried this approach to probability, not only for external applications but also for personal internal reasoning. Again, one can only wonder how reliable the results of all the rigorous deductions can be where the input is so uncertain. It is famous that, as people, logicians are often foolish in their behavior, and that the behavior of the average person based on their intuition is often superior to that of the experts. Indeed, the very jury system of Anglo-Saxon law is a tacit admission that the experts are simply not to be trusted in some of their own areas of competence; that when it comes to innocence vs. guilt the intuition of the jury is preferable to the rigorous knowledge and experience of the judge for delivering justice.

One constantly sees proposals to use probabilistic logic in situations for which it is not suitable. On the other hand, such fields as *operations research* which try to use some common sense, in the hands of sensible people can get valuable results—but in the hands of idiots you get idiocy.

The reliability of the model of probability based on gambling and rigorous mathematics transmits little support to this area of application, but still users try to claim reliability for their results. One needs to be very suspicious and examine repeatedly the robustness at all levels of the discussion before acting on the deductions. Of course, this may lead to "the paralysis of analysis" and no action—which can also be dangerous!

Example 8.8–1   *The Four Liars*

There is a famous problem due to Eddington [E, p. 121]. If A, B, C and D each speak the truth once in three times, independently, and A affirms that B denies that C declares that D is a liar, what is the probability that D was speaking the truth?

Eddington explained his solution as follows:

We do not know that B and C made any relevant statements. For example, if B truthfully denied that C contradicted D, there is no reason to suppose that C affirmed D.

It will be found that the only combinations inconsistant with the data are:

(a) A truths, B lies, C truths, D truths
(b) A truths, B lies, C lies, D lies.

For if A is lying, we do not what B said; and if A and B both truthed, we do not know what C said.

Since (a) and (b) occur respectively twice and eight times out of the 81 occasions, D's 27 truths and 54 lies are reduced to 25 truths and 46 lies. The probabilty is therefore 25/71.

It is not the logic of the deduction (there are still disagreements on it and the concensus may be against Eddington's answer), it is the likelihood of the original assumptions that one wonders about. It is in fact a toy problem having no possible social implications. It is such applications of probability to logical situations that one has to wonder about.

8.9 di Finetti

In the opening of his interesting two volume work on probability di Finetti boldly asserts, (as already noted in the preface of this book), "PROBABILITY DOES NOT EXIST." Of course we need to ask what this provocative statement means, especially the "exists." He also asserts that *independence* does not exist [KS p. 5] and that his probability is the *only* approach [KS p. 1]. He introduces the technical word "exchangeability" which he equates to "equivalence" [KS p. 5]. While perhaps useful it is admittedly nothing new; still it is an idea from which new things may be deduced.

Undergraduates late at night will argue about the existence of the real world, as do some philosophers. But if we take the pragmatic view that how people act is more important in revealing what they believe than are their words, then we see that people act as if the world existed. Those who deny the existence of the real world are still observed to eat when they are hungry. The author is what is commonly called "a naive realist."

Of course all thought exists in the head (so it is widely believed), and we deal with mental models of the world, not with "reality"—whatever reality may mean. But if we assume that there are many kinds of probability, perhaps some exist in the head and some in the "real world." Take, for example, some radioactive material in the lab connected to a Geiger counter and equipment that records each time a radioactive event occurs. When we later analyse the record (using, of course, mental models) we find that the *equipment acted as if there were randomness in the material*. We find, for example, that if we double the amount of material then the number of counts per unit time approximately doubles. Similarly, if we were to build a mechanical roulette wheel spinner and launcher of the pea, or a dice throwing machine, we would find, especially in the first case, that the results were as if they were random, and we would in common scientific practice *infer* that the randomness was "out there in the machine" and not in our head. We can "prove" nothing at all in practice.

To be careful, even Sir Isaac Newton knew that he only gave formulas for *how* gravity worked, and not *why*—"I do not make hypotheses" is what he said in this connection. But the extensive use of the idea of gravity, along with its many, many checks with reality, lead us to say, "Gravity is why things fall." Of course there is no "why" in all of science; we put it there. We see only relationships and we infer the causes; similarly, there are only events and we infer the randomness.

In the above situation of the radioactive material, common sense suggests that we *infer* that the probability exists in the real world and not just in our heads—else we are arguing like undergraduates!

It was by such reasoning, after years of believing in di Finetti's remark cited above, that I came to the conclusion that it is not practical science to adopt his view *exclusively*; that while at times the probability we are talking about exists only in our heads there are times when it is wise to assume that it exists in the real world. It appears to me to be sophistry to claim otherwise.

On di Finetti's side, if I toss a coin and ask you for the probability of "heads" you are apt to say 1/2, but if while it is still in mid air I say to you that it is a two headed coin you will likely change your probability to 1. Nothing in the material world changed, only the state of your information, hence this probability apparently exists in your head. Situations such as horse races are likely to be best seen as having probabilites in your head and not in reality. Thus di Finetti's viewpoint is useful many times. See Example 1.9–8.

To use the same formal apparatus of probability theory that we developed with regard to gambling and such situations for these mental probabilties brings some doubt as to the wisdom of acting as if they had anywhere near the same reliability. Yet, again, what else is there to do in such situations, except to be cautious when acting on the computed result?

8.10 Subjective Probabilities

There are a large number of different subjective probability theories. One theory, for example, believes that a probability is simply "betting odds." The claim is made that by asking you a series of questions on how you would bet in a certain situation, revising the odds at each stage, will lead to your probability of the event. Unfortunately, upon self-examination I find that after a few steps I get stubborn and stick to a single ratio. I simply cannot do what they ask of me, to continually refine my betting odds. Moreover, I seem to come to different probabilities under apparently similar conditions. Thus I do not believe much in the reliability of the probabilities that are obtained this way. See [Ka] for example.

Other subjectivists are more frank about things, and consider that any betting on a horse race is a matter of a local, transient opinion of an individual. Again, these are hardly firm probabilities, though they may be the best that can be obtained. But to then apply an elaborate, rigorous mathematical model of probability, using these estimates, and to then expect the results to be highly reliable seems to me to be folly. There may, however, be some situations where reasonable results can be obtained.

There is at present a strong movement called *Bayesianism*. It is much more connected with statistics than with probability and hence strictly falls outside the field of this book. Still, because of its prominence it is necessary to look briefly at it—though one of its practitioneers, I. J. Good, claims that there are 46,656 varieties of Bayesians!

The Bayesians are the most noticeable of the subjective schools at present. They believe that you do not come to a situation with a completely blank mind, but that you have in your mind a prior distribution for the thing being examined. And in a sense we are all Bayesians. Suppose you and I were to do an experiment to measure the velocity of light, and when our first measurements were made we found them to be quite far from the currently accepted value. We would, almost surely, begin by reexamining the equipment, the reasoning, and the data reduction. But as we made more and more careful adjustments and calibrations we might finally decide that we believed the value we obtained even if it still disagreed with the older accepted value. We would pass, via experience, from one prior distribution of the possible values for the velocity of light to another, but except in carefully rigged situations it is not possible to *quantify* this process with much accuracy.

The Bayesians want us to quantify our prior distribution and use it as a guide for future acceptance (or not) of a result obtained from the experiment. But if the initial distribution of our belief is vague, then one naturally asks,

> *If the prior distribution, at which I am frankly guessing, has little or no effect on the result, then why bother; and if it has a large effect, then since I do not know what I am doing how would I dare act on the conclusions drawn?*

As noted above, in practice we are all Bayesians at heart, but the rigorous accurate quantification of our prior beliefs is usually impossible because they are too vague to handle, so the theory is dangerous to use for important decisions. Furthermore, science has traditionally been "objective" and the Bayesian approach frankly admits "subjectivity" so that different people, doing the same experiments, may well get different results. Yes, the Bayesian approach at times seems reasonable, but not scientific.

Bayesian techniques are often misunderstood by non-professional Bayesians. Their announced posterior probabilites are not thought of as frequencies (how often something will or will not happen), but rather as degrees of belief. But when I ask myself how I will measure the effectiveness of their theory, I naturally turn to testing their predictions against what happens in the real world, how often they are right and how often they are wrong. But in doing so I am adopting the frequency approach which they deny is the meaning of the probabilities they give! One can only wonder about the confidence one can put in their approach. Yes, as we just said, we are all Bayesians at heart, but when it comes to actions in this world their approach, which at times may be the only one available, leaves much to be desired. Still we often must act without good information.

When you must take action and do not have the luxury of firm probabilities then there is little else to appeal to but some form of subjective probability. But again, after using elaborate, complex modelling, to assume that the final results are highly reliable (due to the use of elaborate mathematics) is foolish.

8.11 Fuzzy Probability

In 1965 L. Zadeh introduced the concept of *fuzzy sets*. The concept comes from the simple, but profound, observation that in the real world nothing is exactly what it appears to be, that there is not a sharp dividing line between trees and bushes, that "close enough" is not an exact statement, that a platypus is only partly a mammal, etc. Since then the idea of fuzzy sets has been extended to many fields, including probability, logic and entropy.

The very concept "fuzzy" meets with immediate hostility from the mathematically trained mind that believes that a symbol is a symbol and is exactly that, that each symbol can always be recognized exactly, etc. To be asked to adjust to the idea that nothing is certain, that all things are fuzzy, is rather a lot to expect from them.

Similarly, computer experts like to believe that they live in the classic Laplace universe where if you know exactly the starting conditions and the laws then a sufficiently able mind can calculate the future exactly. The ideal computing machine is a realization of this dream. Although programmers know in their hearts that any large program may have residual "bugs" (more accurately "errors"), still they seem to keep in their minds the concept that a program is exactly right and the machine always does exactly what it is supposed to do. Of course experience has shown again and again that not only are there no exact specifications for most interesting problems, and that large programs are apt to have residual errors in them, but also that the machines themselves are not infallible.

Many people have wanted to believe that fuzziness can be embraced in the concept of classical probability theory. But there is a profound difference between: (1) classical probability theory based on subsets of a sample space, and (2) fuzzy sets where even being in a subset of the sample space is not certain, and knowing which subset you are in is not sure. The first is probability, the second is fuzziness, and they are not equivalent. Quite the contrary, at least some people believe, mathematical logic is merely a part of fuzzy logic!

Once you admit that you cannot know exactly which set you are in, then there are the problems of the laws of noncontradiction and of the excluded middle. Since you can be both in a set and not in the set at the same time, the intersection of the set and its complement is not necessarily 0, nor is their sum necessarily the whole universe.

Is this tremendous mental adjustment worth the effort? It seems that the universe as we know it is indeed fuzzy, and hence it seems that we should give the idea a reasonable hearing—perhaps we might learn something useful! The distaste of accepting that nothing is sure, either in certainty of occurring or in what occurred, is very strong. I think that this accounts for the general rejection ("ignoring" is perhaps a better description) of the ideas, in spite of a small band of eager devotees who have embraced the concept.

8.12 Probability in Science

Scientists use probability theory in many places, and it is natural to wonder which of the many kinds they use. The answer is, of course, different kinds in different places.

Perhaps the classical example is the use of probability in molecular physics, in particular in the theory of perfect gases. Here one of the earliest spectacular results was Maxwell's derivation of the distribution of the velocities of the molecules in a gas. Rather than discuss in detail what Maxwell said, let us look at the essential steps of the argument. We pass over certain simplifying assumptions such as that the molecules are perfect spheres with no intermolecular forces between them, moving independently of each other and independently in the three coordinates.

The argument begins with the assumption of a vector distribution for the velocities. He then argues that in his model between the collisions the velocities of the molecules are constant. Hence he must examine the changes produced by a typical collision. To simplify things in your mind consider the *relative velocity* of one molecule when the other is regarded as stationary. He now assumes a uniform distribution of the axis of the moving molecule with respect to the center of the stationary molecule (there is to be a collision by assumption). He next examines the resulting distribution of velocities after the collision using classical mechanics, and then averages over all intersections and all vector velocities. From equilibrium he assumes that the input distribution of vectors *must be the same* as the resulting distribution after the collision—a self-consistant condition which is widely used in physics. From the resulting functional equation he deduces the velocity distribution. But so far as I have ever seen, though it seems to be obvious, there is no deduction that the uniform distribution of the axes is also true after the collision. Hence the self-consistancy is not quite complete since two probability distributions went in and only one came out!

Now let us analyse a bit of what he was actually assuming. The assumed distribution was certainly not the velocity of a particular molecule at a particular time. Was it the assummed distribution of the many molecules in the volume of gas? That would have given a discrete distribution since there are a finite number of molecules in the assumed volume (which is not really discussed). How does he get to the smooth, analytic distribution he assummed? There are at least two paths. One is to suppose that he is looking at the ensemble of all possible molecules at all possible times, and another is to say that he is merely replacing a discrete distribution by a mathematically more convenient continuous distribution and that the difference must be slight. (Compare Example 5.6–4.)

Must he use arguments that it is uncertainty in initial conditions that produces the probability and that small differences lead almost immediately to large differences after a few collisions? This is one kind of widely used argument in physical applications, but as we have seen he has other arguments he could use which would imply other kinds of probability. In short, we are not sure, unless we examine the matter much more closely, of just what kinds (discrete or continuous density) of probabilities he assumed; but this is not unusual.

Genetics is another field which uses probability widely. The mixing of the genes could be regarded as a purely mechanical system about which we do not have the mechanical details, or it could be regarded as basically probabilistic as in quantum mechanics. Texts are not clear in the details, but then how could one decide which one? Hopefully most of the time the difference would have little effect on the predictions.

Returning to the use of probability in physics, the present *Copenhagen interpretation* of quantum mechanics claims that the probability is essentially present in Nature and is not due to ignorance. This probability, as we will see in the next Section, is a rather different kind from any we have so far discussed. The history of quantum mechanics shows that to this day there is not a uniform agreement on the meaning of their probability; the probability of the single event and the probability as a limiting ratio are both widely held opinions in the field.

In quantum mechanics "probability" is often called the "propensity," apparently to emphasize that it is to be associated with the physical equipment—given an experimental setup C then we have the propensity for A to happen

$$P(A \,|\, C)$$

In quantum mechanics it is necessary to adopt this attitude so that the simultaneous measurement of conjugate variables cannot occur—the act of measurement in quantum mechanics is believed to affect the measured thing, hence using this notation the uncertainty principle cannot be violated since for the other measurement you have

$$P(A \,|\, C')$$

where C' is the corresponding experimental setup. [Ba].

Thus probability in quantum mechanics is different from that in other fields. In quantum mechanics probability is an undefined thing in the sense that it is simply the square of the absolute value of the corresponding wave function and does not have a more elementary description.

In summary, in physics and in science generally it is not always easy to decide just which of the many kinds of probability an author means (because generally speaking the author subconciously believes there is only one kind of probability). The classical mechanical one which supposes that probability arises from uncertainty in the initial conditions is the most popular one, but clearly there are others being used at times.

8.13 Complex Probability

Mathematicians all seem to believe that probability is a real number between 0 and 1. But since 1925 quantum mechanics has used complex numbers as an underlying basis for the representation of probabilities (pre-probabilities if you wish to make the distinction) leading to the probability being the square of the absolute value (modulus) of the complex number and thereby to a real number. Throughout most derivations in quantum mechanics complex numbers are used, and it is only at the last possible moment that the absolute value is taken—indeed, the theory of quantum mechanics was almost completely developed before Born observed that the absolute values squared of the wave functions could be interpreted as probability (propensity) distributions.

Quantum mechanics has been, perhaps, our most successful theory as judged by the number of successful predictions it has made. Typically a wave function is represented in the form

$$\Psi(x) = u(x) + iv(x)$$

where x is a real variable. Ultimately $|u(x)+iv(x)|^2$ becomes the probability. Yet this kind of probability has been generally ignored in probability text books for over 50 years.

It seems unlikely that this is the only application of complex numbers in probability theory (beyond the standard characteristic function representation of distributions which we have not used). We therefore need to examine what is going on. It is useless to examine the history or standard development of quantum mechanics since, as noted above, probability was grafted on (perhaps better, recognized) at a very late stage in the creation of the basic theory.

As a useful analogy we consider Fourier series. It is a linear theory, meaning that the Fourier expansion of a linear combination of two functions is the same linear combination of the separate expansions. However, in applications of the Fourier series the central role is not played by the coefficients of the Fourier series but rather by the power spectrum which is either: (1) the sum of the squares of the two coefficients of the same frequency in the real domain, or (2) the square of the modulus in the case of the complex representation. Thus you *cannot* add the power spectra of two functions and expect to get the power spectrum of the sum, but you can add the coefficients of the corresponding terms in the Fourier representations, and then take the square of the modulus of the resulting complex numbers to get the power spectrum. Thus the physically important things in many of the applications of the theory are quadratic over a linear field of Fourier expansions.

Similarly in quantum mechanics, which according to Feynmann is absolutely linear (not an approximation), the physically important things (probabilities) are quadratic over the linear field of "wave functions" which can be added and subtracted as required. It is the possible cancellations before the squaring of the absolute values that permits the "interference" effects that we see in optics and other physical phenomena.

What else is accomplished by using complex probability beyond the great convenience of a linear theory? If we look at correlation we see that it is related to the phase angle, the angle in the polar form of representing the complex numbers. To illustrate, suppose we represent heads and tails as complex quantities. In the absence, as yet, of formal operators (corresponding to those in quantum mechanics) we guess at the corresponding "wave functions," the pre-probability representations in a vector notation using

$$\Psi = e^{i\phi} = \cos\phi + i\sin\phi$$

and $\Psi\Psi^\star = 1$ (star means complex conjugate) where ϕ is an arbitrary phase angle (as in Fourier series it is the choice of the origin that fixes the phase angles). Now consider a second coin, and suppose that the difference in phase angles is θ with, for convenience only, the original $\phi = 0$.

For a two coin toss we *add* the appropriate components as shown in the following table. The 1/4 in the last column comes from the normalizing process.

TABLE 8.12–1

Outcome	Ψ	$\Psi\Psi^\star$	Probability
HH	$1 + e^{i\theta}$	$2 + 2\cos\theta$	$(1 + \cos\theta)/4$
HT	$1 - e^{i\theta}$	$2 - 2\cos\theta$	$(1 - \cos\theta)/4$
TH	$-1 + e^{i\theta}$	$2 - 2\cos\theta$	$(1 - \cos\theta)/4$
TT	$-1 - e^{i\theta}$	$2 + 2\cos\theta$	$(1 + \cos\theta)/4$

sum = 8

The table shows that for $\theta = 0$, perfect correlation occurs; for $\theta = \pi/2$ there is no correlation and the coins are independent; and for $\theta = \pi$ there a is negative correlation; and finally for $\theta = 3\pi/2$ there is again no correlation. For other angles we have other correlations.

If we wish to follow the quantum mechanical model, then the problem is, in each field of application, to find the appropriate operators (corresponding to the Hamiltonian, etc.) that will generate the "wave functions," the distributions. In the above example I had to guess at a suitable representation. Evidently the problem of finding the proper operators is not trivial, but does suggest that the use of the complex number representation in probability may be useful.

Another approach to complex probability and quantum mechanics has been given by Landé [L]. He gave a derivation of quantum mechanics from some postulates about complex probability. The complaint sometimes made that the derivation does not prove the uniqueness is a trivial point since clearly more postulates (or else slightly altered ones) could probably be found to produce the uniqueness (but then again perhaps the fact that there are equivalent formulations of quantum mechanics means that uniqueness is not necessary or even possible in this field).

8.14 Summary

Chapter 1 noted that, "We speak of the probability of a head turning up on the toss of a coin, the probability that it will rain tomorrow, the probability that the next item coming down a production line is faulty, the probability that someone is telling a lie, the probability of dying from some disease, and even the probability that some theory, say evolution, special relativity, or the "big bang theory," is correct."

We now see that these various probabilities are determined in very different ways with very different philosophies and reliabilities; we see that "probability" is not a single word but is to be understood only in the context of the probability model being assumed.

Philosophers and scientists often claim that for the sake of clarity they want a single word to have a single meaning, yet we have used the word "probability" in many different senses. This is not, in fact, unusal in many fields; we speak of a "line" in both Euclidean and non-Euclidean geometry and mean different things. We could have, of course, used P_1, P_2, ... for the various uses of probability, but found it cumbersome. In practice it often turns out that an author using probability will imply various meanings at various times without apparently being aware of it.

While it appears at first that for the various models the corresponding *mathematical techniques* are the same, we have shown that this is not so. In the infinite discrete sample space we found that it could lead to paradoxes like the St. Petersburg and Castle Point, but if we required the sample space to be generated by a finite state diagram then these paradoxes could not occur. This again raises the question of when the standard mathematical tools can and cannot be used safely *if* we intend to take actions in the real world based on the results. It is not only the modelling of the physical situation that in involved, it is the probability model and the mathematical tools used that are also involved (as well as the logic—see fuzzy sets for example). And as we have carefully argued (Section 7.1 on Entropy) just because you use the same mathematical expressions it does not follow that the interpretations are the same. Probability is indeed an interesting, complicated subject!

Why are there so many different probability theories? I believe that one major reason is the desire that the probability theory the author assumes must support the main ideas of statistics—in short how to get from the definition of probability of a single event to the frequency definition to support statistical techniques. The use of technical words

$$\text{collective} \iff \text{random}$$

$$\text{exchangeability} \iff \text{equivalence}$$

$$\text{propensity} \iff \text{probability}$$

does not really solve much, though they tend to eliminate unintended meanings.

Another major reason is that many applications are quite different from the typical gambling situations where probability theory first arose, and hence need other intellectual tools to handle them.

Why have we glossed over the more subjective types of probability? It is not that we do not use them more or less subconciously (even if we have never heard of the corresponding theory!) rather it is that there are grave doubts that the intuitive elements can be realistically captured into a useful, reliable, reproduceable body of knowledge ready for action in this harsh, real world of science and engineering where important, serious actions involving large sums of money and possibly human lives must be taken based on probability calculations.

The fundamental difficulty with using the subjective types of probability is that success requires mature judgements based on experience, something that cannot be included in a first course in probability. Furthermore, science in the past has highly valued consistency, and subjective probability gives subjective results!

9

Some Limit Theorems

"In the long run we are all dead." Keynes

9.1 Introduction

In the previous several chapters we have been concerned with the initial probabilities and their distributions. In this chapter we will look at how some distributions might arise and hence why it is sometimes reasonable to assume that they are the source of the random events.

The binomial distribution (Section 2.4)

$$b(k; n, p) = C(n, k) p^k q^{n-k} \tag{9.1-1}$$

occurs quite often. We have examined this distribution to the extent of finding the mean and variance (Example 2.7–1)

$$\mu = np$$
$$\sigma^2 = npq \tag{9.1-2}$$

and a few other properties.

We now study this distribution in more detail and answer such questions as how much probability lies in some interval. We are first concerned with the cumulative binomial distribution

$$\sum_{j=0}^{k} b(j; n, p)$$

and then with difference between the function at two different upper limits of this function. Thus we finally have

$$\sum_{j=k_1}^{k_2} b(j; n, p) \qquad (9.1\text{--}3)$$

The analogy with integration is obvious. For n small this sum is easy to compute; for large n we need a way of understanding the size of the sum. Unfortunately there is no simple form for the summation and we are forced, therefore, to find an approximation. We will later be interested in the distribution (see Section 5.8)

$$p(x) = \frac{a^{n+1} x^n}{n!} e^{-ax} \qquad (0 \leq x \leq \infty)$$

Computers can easily compute binomial sums since the terms of the binomial distribution are easily found by recursion (2.4–3), even for fairly large n of the order 100 or more. But if we are to understand what we are doing it is necesary to understand such expressions; hence the necessity of understandable approximations to the cumulative distribution (9.1–3). We have repeatedly observed the simple fact that continuous mathematics is often much easier to carry out than is discrete, hence it should be no surprise that we will use a continuous function

$$\frac{1}{\sqrt{2\pi}} e^{-x^2/2} \qquad (9.1\text{--}4)$$

to approximate the discrete cumulative binomial distribution.

The distribution (9.1–4) is often called *the normal distribution* because it occurs so often (occurs "normally") in probability and statistics. We will, therefore, look a bit at this distribution in the next Section, before we examine how and why it provides a good approximation to the binomial (and other) distributions.

9.2 The Normal Distribution

We are concerned in this section with the function

$$\frac{1}{\sqrt{2\pi}}\, e^{-x^2/2}$$

The factor $1/2$ in the exponent is the common notation. As noted before (Section 5.4) this function cannot be integrated in a finite, closed form, but the definite integral over the whole real line is 1. Thus for the probability density distribution we have

$$\frac{1}{\sqrt{2\pi}} \int_{-\infty}^{\infty} e^{-x^2/2}\, dx = 1$$

The probability density is, of course, just the integrand.

We now examine the shape of the probability density function

$$p(x) = \frac{1}{\sqrt{2\pi}}\, e^{-x^2/2}$$

Neglecting the front coefficient the derivative is

$$-x\, e^{-x^2/2}$$

and is zero only at $x = 0$ with a value $y = 0.3989\ldots$. Of course the derivative is close to zero for $|x|$ very large. The second derivative is

$$(x^2 - 1)\, e^{-x^2/2}$$

and is zero at $x = -1, 1$. These are where the inflection points occur, and the corresponding y values are both $0.2424\ldots$.

To find the first two moments we observe that since the distribution is an even function the mean must be zero. Hence we need only consider the second moment

$$E\{X^2\} = \frac{1}{\sqrt{2\pi}} \int_{-\infty}^{\infty} x^2\, e^{-x^2/2}\, dx$$

Integration by parts, using $U = x$ and $dV = -x\exp(-x^2/2)dx$ leads to the original integral since the integrated part vanishes at both ends. Thus the variance of the distribution is 1, and occurs at exactly the same position x as does the inflection point.

Given the normal distribution

$$p(x) = \frac{1}{\sigma\sqrt{2\pi}}\, e^{-(x-\mu)^2/2\sigma^2}$$

the transformation

$$t = \frac{x - \mu}{\sigma} \qquad \text{or} \qquad x = \sigma t + \mu \qquad (9.2\text{--}1)$$

moves the mean to the origin and provides a scaling so that the variance is 1. Thus this transformation *reduces* a normal distribution with a given mean and variance to a *standard form* of mean zero and unit variance. This makes the study of normal distributions comparatively easy; there is only one standard form and corresponding table of numbers to be considered in practice.

The integral of the normal distribution is often called *the error function*

$$\Phi(x) = \frac{1}{\sqrt{2\pi}} \int_{-\infty}^{x} e^{-t^2/2}\, dt \qquad (9.2\text{--}2)$$

The error function goes from 0 to 1 as x ranges over the whole real line. By symmetry we see that at $x = 0$

$$\Phi(0) = 1/2$$

hence we need a tabluation for only positive arguments, see Table 9.2–1.

The Table 9.2–1 for the normal distribution is easily found by numerical integration (a programmable hand calculator using Simpson's formula with spacing 0.025 produced this table). We append a short table of values for intervals about the origin (9.2–2). This shows that for a normal distribution it is very unlikely that events will fall outside of about 3σ, and 4σ is close to impossible; even 2σ covers more than 95% of the total probability.

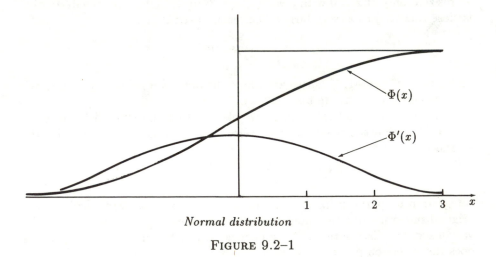

Normal distribution

FIGURE 9.2–1

Figure 9.2–1 shows the error function and the integral—note that as usual the vertical axis is stretched out in comparison with the horizontal axis. Thus the normal curve is much flatter than is generally shown.

TABLE 9.2–1

[Note that $\Phi(-x) = 1 - \Phi(x)$]

x	$\Phi(x)$	$\Phi'(x)$
0.0	0.50000	0.3989
0.1	0.53983	0.3970
0.2	0.57926	0.3910
0.3	0.61791	0.3814
0.4	0.65542	0.3683
0.5	0.69146	0.3521
0.6	0.72575	0.3332
0.7	0.75804	0.3123
0.8	0.78814	0.2897
0.9	0.81594	0.2661
1.0	0.84134	0.2420
1.1	0.86433	0.2179
1.2	0.88493	0.1942
1.3	0.90320	0.1714
1.4	0.91924	0.1497
1.5	0.93319	0.1295
1.6	0.94520	0.1109
1.7	0.95543	0.0940
1.8	0.96407	0.0790
1.9	0.97128	0.0656
2.0	0.97725	0.0540
2.5	0.99379	0.0175
3.0	0.99865	0.0044
3.5	0.99977	0.0009
4.0	0.99997	0.0001

TABLE 9.2–2

$\pm\sigma$	Amount in the double interval
1	0.68268
2	0.95450
3	0.99730
4	0.99994

Example 9.2–1   **The Approximation of a Unimodal Distribution**

A unimodal probability distribution $p(x)$ has the property that $p(x) \geq 0$, hence we can write it as

$$p(x) = \exp\{g(x)\} = e^{g(x)}$$

At the maximum point $x = a$, we have $p'(a) = 0 = g'(a)$, and it is easy to see by differentiating a second time that since $p''(a) \leq 0$ then $g''(a) \leq 0$. The Taylor series of $g(x)$ about the point $x = a$ is, therefore,

$$g(x) = g(a) + (1/2)(x - a)^2 g''(a) + \cdots$$

and hence, using $\exp\{g(a)\} = A$, we have

$$p(x) = \exp\{g(x)\} = A \, e^{\frac{(x-a)^2 g''(a)}{2} + \cdots}$$

For a first approximation about the peak of the unimodal distribution (remember $g''(a) \leq 0$) we have the normal distribution. The approximation is a parabola in the exponent.

Exercises 9.2

9.2–1 Using the integrand $\exp(xt)\exp(-x^2/2)$ find all the moments of the normal distribution.

9.2–2 Approximate the normal distribution by a parabola through the peak and inflection points and find where the parabola becomes negative. Ans. $x = 1.596\ldots$

9.2–3 Approximate sech x by the method of Example 9.2–1.

9.2–4 Show that for $n > 0$ $x^n \exp(-x^2/2) \to 0$ for large x.

9.2–5 Show that for large x, $e^x \exp\{-x^2/2\} \to 0$.

9.3 Binomial Approximation for the Case $p = 1/2$

We now derive the normal approximation for the special case of $p = 1/2$. In this case the ideas of the proof can be clearly seen without the confusion of the more elaborate notation of the general case.

We are interested in approximating the sum of binomial terms in an interval, and for convenience we will use even order only. For n an even number, $n = 2m$, we will have (remember we are doing the simple case $p = 1/2$) the middle term as the maximum of the distribution

$$b(m; 2m, 1/2) = C(2m, n) \left(\frac{1}{2}\right)^m \left(\frac{1}{2}\right)^m = \frac{(2m)!}{m!\, m!\, 2^{2m}}$$

From Stirling's approximation to the factorial (1.A–7) we have

$$b(m; 2m, 1/2) \sim \frac{[(2m)^{2m} e^{-2m} \sqrt{(2\pi)} \sqrt{(2m)}]}{[m^m e^{-m} \sqrt{(2\pi)} \sqrt{(m)}]^2 2^{2m}}$$

Almost everything cancels, and we have left

$$b(m; 2m, 1/2) \sim \frac{1}{\sqrt{\pi m}} = \sqrt{\frac{2}{\pi n}} \qquad (9.3\text{–}1)$$

as the estimate for the largest term in the binomial sequence of values. (We could get bounds if we wished using 1.A–6.)

To get the size of other terms we use k as an index measured (positive or negative) from the maximum, which is at m. Thus we now consider

$$b(m + k; 2m, 1/2) = \frac{(2m!)}{(m + k)!\, (m - k)!\, 2^{2m}}$$

The ratio of this to the middle term gives the relative decrease as we go k terms from the middle

$$\frac{b(m + k; 2m, 1/2)}{b(m; 2m, 1/2)} = \frac{m!\, m!}{(m + k)!\, (m - k)!}$$

$$= \frac{m(m - 1)(m - 2)\ldots(m - k + 1)}{(m + k)(m + k - 1)\ldots(m + 1)}$$

Divide both numerator and denominator by m^k. We have

$$\frac{1\{1 - 1/m\}\{1 - 2/m\}\ldots\{1 - (k - 1)/m\}}{\{1 + 1/m\}\{1 + 2/m\}\ldots\{1 + k/m\}}$$

In (1.B–3) we found a useful bound, and using exactly the same method we can get, (for $y > -1$),

$$\frac{1-y}{1+y} \le e^{-2y}$$

We use this for each pair of factors, and for the unmatched factor $(1 + k/m)$ we use a similarly derived inequality

$$\frac{1}{1+y} \le e^{-y}$$

As a result we have

$$\exp[2\{-1/m - 2/m - \cdots - (k-1)/m\} - \{k/m\}]$$
$$= \exp[-(2/m)\{1 + 2 + \cdots + (k-1)\} - k/m]$$
$$= \exp[-(2/m)\{k(k-1)/2\} - k/m] = \exp[\{-k(k-1) - k\}/m]$$
$$= \exp[-k^2/m]$$

(9.3–2)

Hence, using approximation (9.3–1), we have bound

$$b(m+k, 2m, 1/2) \le \frac{1}{\sqrt{\pi m}} e^{-k^2/m}$$

Since (see the transformation 9.1–2) $\sigma = \sqrt{npq} = \sqrt{n}/2$, if we measure the distance from the mean in units of the square root of variance we will, for this unit, have

$$x = \frac{2k}{\sqrt{2m}} \quad \text{or} \quad k = \frac{\sqrt{2m}}{2} x$$

(9.3–3)

from which the quantity in the exponent, which is what we want, is

$$-k^2/m = -2mx^2/4m = -x^2/2$$

Hence

$$b(m+k; 2m, 1/2) \le \frac{1}{\sqrt{\pi m}} e^{-x^2/2}$$

(9.3–4)

But we are using a spacing of 1 in k, so that from (9.3–3)

$$dk \sim \sqrt{(m/2)}\, dx$$

and we have, finally, the expression, x as in (9.3–3), (and converting the inequality to an approximation since it is close)

$$b(m+k; 2m, 1/2) \sim \frac{1}{\sqrt{2\pi}} e^{-x^2/2}$$

(9.3–5)

is the appropriate density function of the distribution.

Thus we see that the binomial probabilities for $p = 1/2$ fall off from the maximum like $\exp(-x^2/2)$, and that the normal curve has the right shape with the proper front coefficient (so the total probability is 1). The approximation of the binomial distribution by the normal is reasonable for moderate k.

If we picture the original binomial coefficients as rectangles of width 1 centered about their values then we see that the approximating integral should run from 1/2 less than the lowest term to 1/2 above the highest term. See Figure 9.4–1 for $p \neq 1/2$.

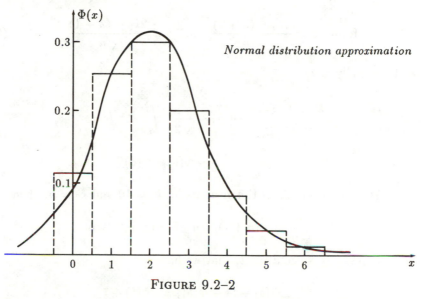

FIGURE 9.2–2

Example 9.3–1   *Binomial Sums*

Sums of binomial terms as given in equation (9.1–3),

$$\sum_{j=k_1}^{k_2} b(j; n, p) = \sum_{j=k_1}^{k_2} C(n, j) p^j q^{n-j}$$

reduce to binomial coefficient sums when we pick $p = 1/2 = q$. We have

$$\frac{1}{2^n} \sum_{j=k_1}^{k_2} C(n, j)$$

For the binomial distribution we have

$$\mu = np = n/2, \quad \sigma^2 = npq = n/2^2$$

hence the transformation to the standard form (9.2–4) gives the range, when we include the extra 1/2 terms,

$$\text{from}\quad a = \frac{k_1 - \frac{1}{2} - \frac{n}{2}}{\sqrt{n}/2} \quad\text{to}\quad b = \frac{k_2 + \frac{1}{2} - \frac{n}{2}}{\sqrt{n}/2}$$

or fixing things up a bit, by multiplying numerator and denominator by 2, we get for the range of integration

$$\text{from}\quad a = \frac{2k_1 - (n+1)}{\sqrt{n}} \quad\text{to}\quad b = \frac{2k_2 - (n-1)}{\sqrt{n}} \tag{9.3–6}$$

and we finally have

$$\frac{1}{2^n} \sum_{j=k_1}^{k_2} C(n,j) \sim \frac{1}{\sqrt{2\pi}} \int_a^b e^{-x^2/2}\, dx \tag{9.3–7}$$

$$\sim \Phi(b) - \Phi(a)$$

As a simple check on this formula let $k_1 = 10$ and $k_2 = n$. then

$$a = -(n+1)/\sqrt{n} \quad b = (n+1)/\sqrt{n}$$

and the result is, since $a = -b$, that we have

$$\Phi(b) - \{1 - \Phi(b)\} = 2\Phi(b) - 1$$

For example, for $b = 4$ we get from Table 9.2–1

$$2(0.99997) - 1 = 0.99994$$

To convert from the $b = 4$ to the corresponding n we have

$$4\sqrt{n} = n + 1 \Rightarrow n^2 - 14n + 1 = 0 \Rightarrow n \sim 14$$

Again, if we pick $n = 9$ we get $b = 10/3 = 3.333\ldots$

$$\text{the estimate is } 2(0.99899) - 1 = 0.99798$$

thus for large values of n the formula for the sum of binomial coefficients is quite accurate.

Example 9.3–2   *Binomial Sums for Nonsymmetric Ranges*

If we use the formula for estimating sums of binomial coefficients for $n = 9$ (where we can check it easily) and pick a short nonsymmetric interval, say $j = 3, 4, 5, 6$ and 7 we get $k_1 = 3$, $k_2 = 7$

$$a = \frac{6 - 10}{3} = \frac{-4}{3} = -1, \quad b = \frac{14 - 8}{3} = \frac{6}{3} = 2$$

Hence $0.97725 - (1 - 0.90855) = 0.88580\ldots$ which is to be compared with $(1/2^9)\{36 + 84 + 126 + 126 + 84\} = 456/512 = 57/64 = 0.8906\ldots$. with an error of $0.0048\ldots$

Exercises 9.3

9.3–1   Use the approximation formula to approximate the middle coefficient of the binomial expansion of order $2n$.

9.3–2   Approximate the $\sum_{j=100}^{200} C(400, j)$

9.3–3   Approximate the $\sum_{j=20}^{30} C(49, j)$.

9.3–4   Approximate the $\sum_{j=5}^{15} C(20, j)$.

9.3–5   Approximate $\sum_{j=6}^{16} C(20, j)$ and check by direct computation.

9.4   Approximations by the Normal Distribution

There are various theorems concerning the approximation of a distribution, usually near its peak, by the normal distribution. Generally the mathematical demonstrations shed little light on why the approximations work. Instead of derivations we will give an intuitive basis for why this happens.

In Example 9.2–1 we showed how in the exponent a single large peak in a distribution can be approximated, near the peak, by a parabola with no first degree term present. Hence this is the normal distribution approximation. In the normal distribution there are three adjustable constants; the front coefficient, (usually chosen so that the total area is 1), the location of the maximum of the approximation for symmetric distributions (which is a shift in the coordinate system), and finally the square root of the variance (which provides a scale factor for the spread of the approximation).

For a nonsymmetric distribution the problem immediately arises, what kind of an approximation should we use. The Taylor series approach (even up in the exponent) makes the fit very good in the immediate region of the peak, but generally the quality of the fit falls off rapidly as you go away from the peak; it is good *locally* but not *globally*.

A more commonly used *global* method of fitting one curve to another is to match the moments of the two distributions, and for the normal distribution this means the first three moments, 0, 1 and 2. The zeroth moment means we will have exactly unit area under the approximating curve, the matching of the first moments forces us to have the means of the two distributions to be the same (we are abandoning the match at the peak), and the second moment will give matching spreads of the distributions. This is a *global* rather than a *local* approximation.

Other methods are available; for example picking the curve with the least squares difference. With modern computers this is quite reasonable to try. However, the method of matching moments is the traditional and usually fairly effective one, hence we will use it.

The choice of the normal distribution means that when we have a non-symmetric distribution there will be noticable errors near the peak. Furthermore, for the nonsymmetric distribution (5.8–1)

$$p(x) = \frac{x^n}{n!} e^{-x} \qquad (x \geq 0)$$

the normal approximation goes to minus infinity but this distribution is zero on the negative half of the real line. The errors in the tails are often of less importance than they seem at first glance, since the actual area under them may be very small indeed. The probability of being at most $k\sigma$ away is given in the Table 9.2–2.

Example 9.4–1   *A Normal Approximation to a Skewed Distribution*

Approximate the distribution

$$p(x) = \frac{x^3}{3!} e^{-x} \qquad (x \geq 0)$$

by the normal distribution.

By integration we have the first three moments of this distribution are 1, 4 and 20, hence the mean is 4 and the variance is $20 - 4^2 = 4$. The transformation to the standard variable is, therefore,

$$\frac{x - 4}{2} = t \qquad \text{and} \qquad dx = 2\,dt$$

and we have the following table comparing the original and approximate distributions.

TABLE 9.4–1

x	$2\left(\frac{x^3}{3!}\right)e^{-x}$	t	$\Phi'(t)$
0.0	0.0000	−2	0.0540
0.5	0.0032	−1.75	0.0836
1.0	0.1226	−1.50	0.1295
1.5	0.2510	−1.25	0.1826
2.0	0.3609	−1.00	0.2420
2.5	0.4275	−0.75	0.3011
3.0	0.4480	−0.50	0.3521
3.5	0.4316	−0.25	0.3867
4.0	0.3907	0.00	0.3989
4.5	0.3374	0.25	0.3867
5.0	0.2807	0.50	0.3521
5.5	0.2266	0.75	0.3011
6.0	0.1785	1.00	0.2420
6.5	0.1376	1.25	0.1826
7.0	0.1043	1.50	0.1295
7.5	0.0778	1.75	0.0863
8.0	0.0573	2.00	0.0540
8.5	0.0417	2.25	0.0317
9.0	0.0300	2.50	0.0175
9.5	0.0241	2.75	0.0091
10.0	0.0151	3.00	0.0044

We see that the approximation is reasonably good.

Example 9.4–2   *Approximation to a Binomial Distribution*

Let us return to the approximation of the discrete binomial distribution by the normal distribution. In Figure 9.2–2 we see such an approximation for $p = 1/5$ and $n = 10$. In the figure we see, as it were, the normal distribution struggling to fit the skewed binomial, and all things considered it is doing a good job. This simple example shows the quality of fit one can expect, and the more skewed the original distribution is the more trouble the normal will have in approximating it.

To find the actual normal approximation parameters, as stated above we match the first three moments. The zeroth moment is automatically matched by the standard front coefficient of the normal distribution. We have that the mean of the binomial is np (see Example 2.7–1) and the variance is npq. We have, therefore, the $x_0 = np = 10/5 = 2$, and the $\sigma = \sqrt{(npq)} = \{(10)(1/5)(4/5)\}^{1/2} = \sqrt{1.6}$. Thus we introduce *the reduced variable*

$$\frac{k - np}{\sigma} = \frac{k - 2}{\sqrt{1.6}}$$

and use it to consult the normal distribution.

To summarize, for the normal approximation to a discrete distribution we take the range and add 1/2 to the upper limit and subtract 1/2 from the lower limit (to allow for the rectangular blocks we used in Figure 9.4-1), shift the center by subtracting the mean, and then scale by dividing by r. As a result to get the approximation to the sum we have to look up the two limits in the table and take the difference.

Exercises 9.4

9.4–1 Approximate $\sum_{k=20}^{80} C(100, k)(1/3)^k (2/3)^{n-k}$.

9.4–2 Approximate $\{x^4/4!\}\exp(-x)$.

9.4–3 Approximate $b(k; n, p)$ generally.

9.5 Another Derivation of the Normal Distribution

Another source of the normal distribution is that the limit of a sum of a large number of random variables is the normal distribution. Again we will not "prove" it mathematically, rather we will indicate why it happens.

Suppose we think of a bounded flat distribution, such as arises in round-off theory when we look at the part that is omitted by the rounding of a number. Theory and practice indicate that this distribution is flat from $-1/2$ to $1/2$. We plan to look at the sum of three such random numbers drawn from this distribution. Since the variance of this distribution is 1/12 the sum of the three independent random numbers will have variance $3/12 = 1/4$.

Because we are going to compare it with the standard normal distribution which has a variance of 1 we are led to consider picking the three random numbers from the distribution of double the length

$$p(x) = 1/2, \quad (-1 \le x \le 1) \qquad \text{and 0 elsewhere.}$$

For this distribution the mean is zero and the variance is 1/3 so that the variance of the sum of three independent samples is 1.

To get the sum of independent random numbers we convolve their distributions. It is easy to see that the convolution of two of these $p(x)$ is a triangle with the base reaching from -2 to 2, and the peak of $y = 1/2$ at $x = 0$. When we convolve this with the rectangular distribution $p(x)$ using simple, though tedious, calculus, we get a quadratic from -3 to -1, another quadratic from -1 to 1, and finally a corresponding quadratic from 1 to 3. The following

TABLE 9.5–1

x	Calculated	Normal approx.	True–approx.
0.0	.3750	.3989	−.0239
0.2	.3700	.3910	−.0210
0.4	.3550	.3683	−.0133
0.6	.3300	.3332	−.0032
0.8	.2950	.2897	+.0053
1.0	.2500	.2420	+.0080
1.2	.2025	.1942	+.0083
1.4	.1600	.1497	+.0103
1.6	.1225	.1109	+.0116
1.8	.0900	.0790	+.0110
2.0	.0625	.0540	+.0085
2.2	.0400	.0355	+.0045
2.4	.0225	.0224	+.0001
2.6	.0100	.0136	−.0036
2.8	.0025	.0079	−.0054
3.0	.0000	.0044	−.0044

Table 9.5–1 gives the result and compares it with the corresponding normal distribution. We compare the density distributions (rather than the cumulative) and only for the positive half since they are both symmetric with respect to the origin.

We see from this Table that the normal distribution, even for only three random numbers from a flat distribution, is a remarkably good approximation. The biggest error is less than $2\frac{1}{2}\%$ and occurs at $x = 0$ where the function is largest, see Figure 9.5–1.

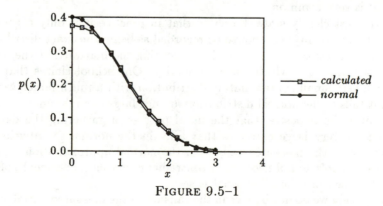

FIGURE 9.5–1

As you think about convolutions you soon realize that a convolution is a *smoothing* process, and that the exact flatness of the original distribution is not necessary. Any reasonable distribution of finite range would do, but certainly not the "spike" delta function that is used to denote all of the distribution at a point.

Upon further thought you see that it is not necessary that each random number entering into the sum come from the same distribution; they might come from comparable distributions so long as the contribution from any one tends to decrease to zero as the number of random numbers approaches infinity. Still further thought suggests that the original distribution need not be finite in range, though again some limitations must be applied on the source distributions.

Thus we see that the sum of a large number of small random effects often tends to approach a normal distribution. In practice we seldom can estimate the number of independent contributions entering into the total effect, nor do we often know much about their individual distributions, so the exact mathematical theorem, while suggestive, is rarely rigorously applicable.

If we have discrete distributions, say for example

$$\frac{\delta(x - 1/2) + \delta(x + 1/2)}{2}$$

where $d(x)$ is the usual delta function (a function with a single peak of no width but with total area 1), then the first convolution will give a distribution of

$$1/4, \quad 1/2, \quad 1/4$$

and the following ones will generate the corresponding binomial coefficients. But we saw that if we regard the spikes as rectangles then we will get a nice normal approximation in the limit.

From all of the above we see, finally, why the normal distribution arises so often in probability theory. As its name suggests (there are other distributions), it is very common.

We must clearly state, however, that in practice it is only a good approximation to reality; it is not to be regarded as being the exact distribution. There are various schools of thought concerning the matter that the normal distribution has tails which reach to infinity. One school claims that often there is a very small contaminating distribution with a much larger variance and that tails of the normal distribution are not large enough; another school claims quite the opposite, that the usual process of gathering the data has censored the very large extremes that arise in the normal distribution and one has less in the measured tails than the normal approximation suggests. These two schools do *not* necessarily contradict each other; one speaks of what should be seen, and the other what is seen.

Certainly we do not believe in the tails when the measured variable must not be negative (the distribution $p(x) = (x^n/n!)\exp(-x)$, $(0 \leq x < \infty)$, for example) nor when the original distribution we are approximating is skewed about its mode. And if we are dealing with a distribution that is not unimodal we can use the normal approximation only near a peak (Example 9.2–1). Still, the normal distribution occurs frequently and can be expected to arise in many situations.

Exercises 9.5

9.5–1 Find the approximation polynomials used in Table 9.4–1.

9.5–2 Find the approximation for the sum of four random numbers from the flat distribution $(-1, 1)$.

9.5–3 Do a monte Carlo simulation of: (the sum of 12 random numbers from the flat distribution)—6. Compare with the normal distribution.

9.5–4 Convolve the distribution $p(x) = (3/4)(1 - x^2)$, $|x| \leq 1$, with itself and compare with the normal distribution.

9.6 Random Times

As we did in the previous Section, we will indicate, not prove, suggestive theorems. The lack of rigor between the theory and the application is so large that the rigor of derivation is often misleading.

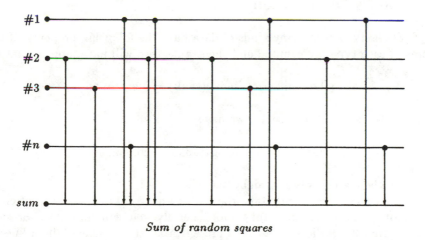

Sum of random squares

FIGURE 9.6–1

Suppose we have a number of *independent* processes each having essentially random times of occuring (though not perfectly random perhaps). See Figure 9.6–1 where the events of each imagined process are schematically indicated. If we combine these events into a single stream (project the events onto a common line as in the Figure), then the intervals between events will often be close and only occasionally be far apart. Of course it is easy to imagine counter examples which will require restrictions on the applicability of the result, but the underlying idea is simple; a large number of independent sources of events, when combined, tends rapidly to a the exponential distribution for the interarrival times of the events. In many respects it resembles the earlier claim that enough single events when added approach the normal distribution. In both cases it is indicative, and we rarely have reality so nice as to know the details, or how many independent sources there are. Thus it is the intuitive, rather than the rigorous, theorems that are used in practice.

We have repeatedly warned that rigor in one part of an application does not cover the lack of knowledge of the hypotheses of the rigorous theory. We must acknowledge, however, that in practice, in corporations, in science, in engineering, and in industry generally, this mistake is repeatedly made; a lot of different sources are combined and then the rigorously found bounds of uncertainty on many of the parts are taken to also cover the parts that are wild guesses at best. We have preferred to stress that in applications of the limit theorems rarely do we have much more than intuition to guide us. Radioactive decay and calls on a central office are examples where the limit theorems give good results.

9.7 Zipf's Distribution

Zipf [Z] observed that many kinds of data have the following property: if the items of each type are counted and then ranked we will have, approximately,

$$\text{(the number of items)(rank)} = \text{(constant)}$$

Taking logs (say to the base 10) we have

$$\log(\text{number}) + \log(\text{rank}) = \text{constant}$$

which in the log units is a straight line of slope -1.

For example, one may take the U.S. Census data on the size of cities and plot log size vs. log rank, and generally you will find that for each census, from the earliest data to the present, that the numbers fall on lines of slope -1. Of course the largest cities will not fall exactly on the line; there will be a period when the city of Philadelphia is being surpassed by the city of Chicago as the city of second rank, and in turn Chicago is displaced later by Los Angles. But once past the first few cities, you will find that the tenth, hundredth, thousandth, and ten thousandth sized cities fall remarkably on the stated lines.

Similarly, it has been shown that the same rule applies to the usage of words in a book. Indeed, Zipf's book is filled with many, quite diverse examples of this rule.

It must, however, be realized that the rule implies a basic unit of size below which we do not go. The population of cities goes by units (people), as does the number of times a given word appears in a text. If we went to fractions then since the harmonic series

$$\sum_{j=1}^{\infty} \frac{1}{j}$$

diverges the total population would be infinite! Thus the analytic use of the continuous function $y = 1/x$ to represent the curve must be truncated at some finite point—the point where the area under the curve equals the total population of the study.

The rule has been applied to so many different situations with such remarkable results that a theoretical basis has been long sought. Zipf tried to base it on the natural economy of the situation, following the then popular least action, least time, and least work laws that were of great importance in the growing field of mechanics. Due to the diversity of applications it seems to this author that no single derivation will be applicable for all situations, *unless* it is based on some statistical property of ranking things. It is clear from the nature of the the data that the curve must be monotonely decreasing, and there have been applications that suggest that the slope in not always exactly −1, but it is generally close.

The application to cities we mentioned above means that in some sense the smaller sized cities "feel" the presence of all the other cities in the political unit being studied, that somehow they adjust their size to the rest of the entire unit —and when you do not find the straight line it is inferred, with some justice, that you do not have the proper large unit in which you are counting. Thus Zipf's law is a global measure of internal relationships within the general population being counted. But without a derivation from some basic principles one can only wonder what the deeper meaning of Zipf's rule is.

9.8 Summary

We have given rather heuristic derivations of a few limiting distributions. The usual rigorous derivations are apt do deceive the user into thinking that Nature supplies such limiting distributions, rather than realizing that they are convenient approximations to be used in estimating things. The normal approximation to bionomial sums is useful mainly for thinking about situations (modern computers can compute most binomial sums rapidly and accurately). Because Zipf's distribution has at present no good theoretical basis it is generally ignored in text books, but its prevalence required us to at least mention it.

10

An Essay On Simulation

Simulation is better than reality!

10.1 Introduction

We have already done a number of simulations to check our theoretical deductions. They are also very useful when we are not able to compute the answer to the problem posed, and even more useful when in some situations it is very hard to cast the problem into a clean mathematical form suitable for mathematical solution.

The development and spread of computers has greatly increased the number of simulations done, but the idea is not new. The Buffon (1707–1783) needle estimate of π is an old example. In the late '20s the Bell Telephone laboratories ran simulations of the behavior of proposed central office switching machines using a combination of girls with desk calculators plus some simple mechanical gear and random number sources. To solve a number of problems in particle physics Fermi did a some simulations back in the late '30s. Modern computers have popularized the idea and made many simulations relatively painless. This chapter merely indicates some of the possibilities of simulation as related to probability problems.

It is not possible to give a precise definition of what a simulation is. Simulations on computers grade off into simulations in laboratories, and on in to classical laboratory experiments, which in their turn were often regarded as simulations of what would happen in the field. Thus we have to leave the concept undefined, but what we mean is fairly clear—it is the limits of the idea of a simulation that apparently cannot be defined sharply. In a sense machine simulations, (since they are usually cheaper than laboratory experiments and can often do what no real experiment can do), have brought us back to the Medieval scholastic approach of deciding by mere thinking what will happen in a given situation without looking at reality (Galileo, the founder of modern science, advocated looking at reality and opposed this conceptual approach, though he was guilty of using it himself many times.)

10.2 Simulations for Checking Purposes

The main use we have made of simulations has been to check results we computed. Such simulations are valuable, but have their disadvantages. For example, suppose the simulation and the theory disagree; what to do? The difference may be due to:

(1) an error (bug) in the coding,

(2) an error in the plan of the program so that it does not represent the problem,

(3) a misinterpretation, for example you may have computed the median value and be comparing it with the expected value,

(4) the theoretical result may be wrong,

(5) you may be merely facing a sampling fluctuation.

As you try to find the cause of the differences between the theoretical value and the simulation you may need to repeat the simulation a number of times, hence you must be able to use exactly the same run of random numbers again and again—something that you need to think about *before* starting the simulation!

Sampling fluctuations, for reasonably sized samples, will not bother you much since you can often increase the size of the sample until it is very unlikely to be the cause of the difference. But persistant small differences between the theory and the sample are troublesome, and should be pursued until you are satisfied as to the reason (it may be an approximation in the theory).

10.3 When You Cannot Compute the Result

Often you cannot analytically compute the results you want from the model, either because the mathematics at your command are inadequate, or because you simply cannot formulate the problem in a sufficiently mathematical form to handle it. Then, of course, you have only simulations and experiments to guide you. When experiment and simulation differ, that can be annoying but can also be a possibility for finding new things that the experimentalist never thought about. The difference should not be swept under the rug and explained away as a sampling fluctuation; it is an opportunity to learn new things.

It has been said, "The only good simulation is a dead simulation." and in the author's experience one result obtained at great labor (in those days) suggested in a few minutes the right way to think about the problem and hence how to derive the desired equations for the general case. Hence a simulation may show you how to solve a problem analytically.

Example 10.3–1   *Random Triangles in a Circle*

What is the probable area of a random triangle inside a unit circle?

While with great effort this can be solved in closed form to get the result $35/(48\pi) = 0.2321009\ldots$ it is easy to simulate and get an approximate answer. To get a random point in the circle you pick a pair of random numbers in the range 0 to 1 and transform them to the range -1 to $+1$ by the simple transformation

$$x' = -1 + 2x$$

You then calculate the sum of the squares to find if the point is inside the unit circle. About $1 - \pi/4$ of the time you will be outside and have to compute a new point inside the circle. Thus to get a random triangle you will need about 4 trial points per triangle. You can now compute the area by the standard formula

$$A = \frac{1}{2} \begin{vmatrix} 1 & x_1 & y_1 \\ 1 & x_2 & y_2 \\ 1 & x_3 & y_3 \end{vmatrix}$$

and take the absolute value of this result to get a positive area. You are not going to get the exact answer by this simulation, but it is likely that you can get a good enough result for most engineering purposes. It is certainly easier and faster than the theoretical computation, but for a theoretician less interesting.

Exercises 10.3

10.3–1   Simulate finding the area of the triangle in Example 10.3–1.

10.3–2   Simulate the area of a random triangle in a square.

10.3–3   Compute and simulate the probability of a unit circle falling at random completely in a square of side 2 if the center is randomly chosen in the square.

10.3–4   If two points fall in the unit square and determine a line, find the probable distance of a third random point from the line. [*Hint:* use the normal form of the line.]

10.4 Direct Simulations

Most simulations are devoted to comparing alternate designs so that a good design can be found. Hence it is often the *differences* between simulations more than the actual numbers that matter.

For purposes of discussion we shall assume that we are concerned with the possible "blocking of calls" in a proposed central office. Suppose we have 100 recorded cases of when a phone call is placed and how long it lasted. To see how things would go with the proposed central office we could simply use these 100 cases and see, trying one call after the other, what would happen (with of course an inital state of the central office also assumed). From the simulation we could assert that had the central office been in that state at the start and seen that previously recorded set of 100 calls then what we simulated would have been what would have been seen (assuming of course that the simulation is accurate). By trying the same sequence of calls on the alternate designs we could evaluate which design would have done better (in terms of the measure of success we have decided on) for that particular sequence of calls.

In the beginning of the design, when a reasonable fraction of proposed designs will be better than the current one, this simple proceedure will tend to approach a good design. But when you are near an optimum then most variations of the design will be poorer than the current one, and due to the chance fluctuations of the simulation (the actual sample you used) on a single step of the optimization you would often choose an inferior design over a better one. Hence as you near an optimal design you need to increase the number of calls simulated before deciding which design is preferable.

One way to increase the number of trial calls is simply run through the list several times, and since almost surely the inital state of the central office would be different for each sequence of the 100 trials you will get new information about the behavior of the design.

This exact repetition of the same sequence of calls leaves one with an uneasy feeling that some peculiar feature of the sequence might just happen to favor one possible design over the others. Hence you think that maybe you should select the next call at random from the list of 100 recorded calls.

This raises the question, "Sampling with or without replacement?" especially if you intend to run through say 4 or 5 hundred calls. We showed in Example 3.6–3 that without replacement then a sequence of 100 random choices would get every recorded call, but that with replacement you could expect to miss about $1/e$ of the original recorded calls. Of course on the second or still later repetition of 100 calls you would get some of them. Still, using sampling without replacement each recorded call would enter equally in the final simulation, while with replacement there would a fair amount of unequal representation in the number of times various apparently equally likely original calls were represented in the simulation. Of course you can say that due to the random sampling the result is unbiased, but unequal representation of the original data in the simulation leaves one feeling dubious about the wisdom of sampling with replacement.

It should be noted that obviously the same sequence of random choices should be used in each comparative trial since this will reduce the variability in the comparisons (without improving the reliability of any one simulation).

Exercises 10.4

10.4–1   Show that $1/e^2 \sim 0.135$ is a good estimate for the number of missed items in taking 200 samples from 100 items when sampling with replacement.

10.4–2   Similarly, that for 300 samples you will miss about 5%.

10.4–3   Show how to randomly sample without replacement a list of n items.

10.5   The Use of Some Modeling

It may be that from other sources you have reason to believe that the inter-arrivals times of the calls are closely modeled by an appropriate exponential distribution. If this assumption is correct then you can use the recorded data only for the length of calls and use random times from the appropriate exponential distribution to simulate the times the calls occur.

If the assumption of the appropriate exponential distribution is indeed valid then, since you have put in more information, you will get better results. But if it is not an accurate representation then you have done yourself harm. It is the same situation as in the use of nonparametric or parametric statistics; nonparametric statistics gives poorer bounds than parametric statistics, but if the assumed model is seriously wrong then the parametric statistics result is almost surely worse! Similarly, when you model the time of occurence and the duration of the calls, then insofar as you get them correct the simulation will be better, and insofar as you get them wrong you will get worse answers, though the statistical estimates will indicate better results!

There are, however, other aspects that need to be discussed. In all the above there is clearly the assumption that the future will be like the past. Of course if you have knowledge of trends these can be incorporated into the simulation—at the risk of being worse off if your trend assumptions are bad. There is, always, the assumption that the sample you originally have is typical and not peculiar, so there is always the problem of the original data being misleading.

Another aspect, greatly neglected by the experts, is the question of the believability of the simulation by those who have the power of decision. In an inventory simulation, once I realized that the vice presidents who would make the decision were quite intelligent but wary of hi falutin' mathematics (perhaps justly so!), I therefore used only the simplest modeling and I could then state, "If you had used the proposed inventory rules at the time the data was gathered then you would have seen the results we exhibited (with the quibble that some few events were missed and some few would not have occurred)." This simulation carried enough conviction to produce further work on their part.

Therefore, before you start any simulation thought should be given to this point—for whom is the simulation being done? Who has the power of decision? If they are not willing to act on your simulation then for whom is it being done? Yourself? (The latter may well be true!) The crude simulation mentioned above produced action, especially as we could (and did) turn over the crude method to their people to be run on their machines. We did not remain in the loop of computing beyond teaching them how to run the simulation, what it meant, and then stood beside them, as it were, whenever they asked for help. Thus the use of elaborate simulation models to meet ideal conditions, rather than practical needs, is often foolish. On the other hand, if the deciders lack confidence in their own judgements then they are apt to want every known effect included, whether or not it is significant. How you go about a simulation depends somewhat on your understanding of the minds of the consumers.

It is remarkable how many simple simulations can be done with merely paper and pencil, and at most a hand calculator. Such simulations often reveal "what is going on" so that you can then approach the problem with the proper analytical tools and get solutions which are "understandable by the human mind" rather than get tables of numbers and pictures of curves. Indeed, after every simulation the results should be closely inspected in the hopes of gaining insight and hence long term progress.

It should be noted that there are often situations in which it is difficult to describe the system under study in standard mathematical notation. In many of the above kinds of simulations all one needs are the logical relationships between the parts, and when and where things happen to the system. This is a great advantage in ill-explored areas where the backlog of mathematical formulations and results has not yet been built up.

10.6   Thought Simulations

We have already illustrated in the text a number of thought (gedanken) simulations. For example, in the gold and silver coin problem, Example 1.9–1 and following, we imagined several different simulations, noting which cases were eliminated, how many trials there were, and how many were successes. From this we understood why the analytical results were reasonable. Had we not had the analytical result the corresponding thinking might well have led us to a good estimate of the result, and likely indicated how to get the analytical answer.

This method of thought experiments is very useful when not much is known or understood; you merely imagine doing the simulation, and then by mentally watching it you often see the underlying features of the problem. It may, at times, be worth thinking of the detailed programming of the proposed simulation on some computer even if you have no intention of doing it—the act of programming, with its demands on explicitly describing all the details, often clarifies in your mind the muddled situation you started with.

10.7   Monte Carlo Methods

At present the name of Monte Carlo (which comes from the name of the famous gambling place) is used to refer to any simulation that uses random numbers. Originally it was used to cover those problems which were deterministic and where there was a random problem which would lead to the same set of equations. Thus solving the random problem gave an estimate of the solution of the corresponding deterministic problem. Buffon's needle is an example of replacing the computation of a perfectly definite number π with the simulation of a random situation that has the same solution.

If you are going to do a large Monte Carlo problem then it is wise to look at the topic of "swindles," the deliberate introduction of negative correlations to reduce the variability. For example, in the Buffon needle problem if we replace the needle by a cross of two needles, then on any one toss if one bar of the cross does not intersect a line then it is highly likely that the other bar will produce an intersection. In this fashion we see the reduction of the variance and hence the reduction in the number of trials (for a given reliability) used to estimate the number.

10.8 Some Simple Distributions

One usually has a random number generator that has been supplied, either on the hand calculator you own, on the personal computer you have, or on the large central computing facility. The random number generator usually produces 1/4 of all the possible numbers that can be represented in the format used, and are uniformly distributed in the range of 0 to 1, [H, p. 138]. The generator goes through the whole set and then exactly repeats the sequence. Thus one may regard them as doing a random sampling without replacement. If you were to remove the lower digits of the numbers from the generator then you would have each number in the sequence repeated exactly the same number of times when you got around the whole loop. If you want truly random numbers, meaning among other things sampling with replacement, then the random number generator does not deliver what is wanted.

After years of effort on the problem there is still no completely satisfactory random number generator, and unless you are willing to devote a very large amount of effort you will probably settle for the generator supplied. You are advised, however, because there are still a number of very bad generators in use, to make a few of your own tests that seem to be appropriate for the planned use before using it.

We often need random samples from distributions other than the flat distribution which the random number generator supplies. To get numbers from an other distribution, $f(y)$, we start with the uniform distribution between 0 and 1 and equate the corresponding points on their respective *cumulative* distributions, that is we set

$$\int_0^x 1\, dx = \int_{-\infty}^y f(t)\, dt = F(y)$$

Thus we have the equation

$$F(y) = x$$

If we can analytically invert the function $F(y)$ we have the corresponding transformation

$$y = F^{(-1)}(x)$$

to make on each random number x from the uniform generator.

Example 10.8–1   *The Exponential Distribution*

Suppose we want random numbers from the distribution $f(y) = \exp(-y)$. As above we get

$$x = \int_0^y e^{-t}\, dt = 1 - e^{-y}$$

Solving for y we get

$$y = -ln(1-x)$$

but since x is uniform from 0 to 1 we can replace $1 - x$ by x to get

$$y = -ln\, x$$

as the transformation.

You cannot invert analytically the cumulative distribution of the normal distribution, so we have to resort to some other approach.

Example 10.8–2   *The Normal Distribution*

To get random numbers from a normal distribution you can add 12 numbers from the flat generator and subtract 6 (see Example 5.6–2). The mean of the random number generator is 1/2 so that the subtraction of 6 from the sum gives a mean of exactly 0. Since the variance of the original flat distribution is 1/12 the 12 independent numbers give a distribution with variance exactly 1.

The above could be expensive of machine time, and if you are going to use a large amount of machine time generating the numbers from a normal distribution in your simulation then you need to look farther into the known methods of getting them cheaper—however, if you try to develop your own it may well take more machine time to test out your ideas than you will probably save!

In table 9.4–1 we saw that even the sum of 3 random numbers from the flat random number generator closely approximated the normal distribution; the sum of 12 naturally does much better. But of course all your numbers will be within $\pm 6\sigma$. Table 9.2–2 shows that the tails of the normal distribution are negligible at 4σ, and they are infinitessimal at 6σ since we have about 2×10^{-9} outside the range and we can safely ignore the tails *except* in simulations which depend heavily on the effects of the tails.

Example 10.8–3   *The Reciprocal Distribution*

To get mantissas for simulating floating point numbers from the reciprocal distribution we rely on the observation that succesive products from a flat distribution, (provided we shift off any leading digits that are 0, rapidly approaches the reciprocal distribution. That is, we use the simple formula, where the x_i are the random numbers from the uniform distribution and the Y_i are those we want,

$$Y_n = Y_{n-1} x_n \qquad \text{(shift)}$$

with shifting to remove any leading zeros in the products. Tests show that this is remarkably effective, [H; p. 144].

This is not a book on statistics; in such books you can often find methods of generating random numbers from all kinds of distributions, so we will not go on developing them here. It is sufficient to say that most distributions have fairly simple methods of generating random numbers from them.

10.9   Notes on Programming Many Simulations

If you plan to do many different simulations in the course of your career, it is wise to think about taking the time, as you do the individual ones, to build the tools for the class of simulations you expect to meet in the future. Of course you want to get the current simulation done, and to build a general purpose subroutine to do various parts seems to be a waste of time (now), but a good workman builds the tools to do his trade. No small set of subroutines can be given since individuals tend to work in various areas and therefore have different needs. It is foolish to try to build the whole library before you ever start; it is equally foolish never to build any parts at all. Each person should think through their needs. There are, of course, packages that are designed for such purposes, but again the effort to get one, then domesticate it on your particular machine, is apt to be much more than most people initially think it will be. The library programs will, generally speaking, be better than those you build for yourself, but building them yourself means that you will understand their peculiarities, and not be misled by some odd feature you never thought about when doing the simulation—thus potentially vitiating the whole effort with no warning!

10.10 Summary

Simulations, now that computers are widely available, are very practical, especially in the field of probability, and their value is apt to be overlooked by theoreticians as being beneath their dignity. But amateurs are apt to overestimate their value and reliability.

Crude simulations done *early* in the problem can give valuable guidance towards how to solve the problem; accurate simulations are often very expensive (but running a personal computer all night does not use a lot of electricity nor produce much wear and tear on the computer).

Even after an analytic result is obtained, the numerical simulation may shed light on *why* the result is the way it is—in probability problems we often have a poor intuition so that the details of the simulation may illuminate many a dark corner of our experience.

References

[Ba] Ballentine, L. E. Probability Theory in Quantum Mechanics, Am. J. Phys. 54 (10) October 1986, pps. 883–889

[B] Bishop, Errett. Foundations of Constructive Analysis, McGraw-Hill Co. 1967

[C] Cacoullos, T. Exercises in Probability, Springer-Verlag, 1989

[Ga] Gatlin, L.L. Information Theory and the Living System, Columbia Uni. Press, 1972

[dF] di Finetti, B. Theory of Probability, John Wiley and Sons, N.Y. 1970

[D] Diner, S., Fargue, D., Lochak, G., and Selleri, F., The Wave-Particle Dualism, D. Reidel Pub. Co. 1984

[Ed] Edwards, A. W. F., Likelihood, Cambridge Uni. Press, 1972

[Ef] Efron, B. Why Isn't Everyone a Bayesian?, Am. Statistician, Vol. 40, #1, Feb. 1986, p. 1–5.

[E] Ekeland, I. Mathematic and the Unexpected. U. of Chicago Press, 1988

[F] Feller, William An Introduction to Probabilisty Theory and its Applications, John Wiley and Sons, Third Ed. 1968

[G] Gudder, S.P. Quantum Probability, Academic Press, Inc. 1988

[H] Hamming, R. W. Numerical Methods for Scientists and Engineers, 2nd Ed. Dover Publications, 1986

[K] Kac, Mark. Statistical Independence in Probability, Analysis and Number Theory, Carus Monograph, Math. Soc. of America

[Ka] Kaplin, Mark. Bayesianism Without the Black Box. Philosophy of Science, Mar. 1989, p. 48–69.

[Ke] Keynes, J. M. A Treatise on Probability, MacMillan, 1921

[Kh] Khinchine, A. I. Mathematical Foundations of Information Theory, Chelsea 1957

[Kr] Kruger, L. et al. The Probabilist Revolution, 2 Vol. MIT Press, 1987

[K–S] Exchangability in Probability and Statistics, Koch, G. and Spizzichini, F. North Holland, 1982

[L] Landé, A., New Foundations of Quantum Mechanics, Cambridge Uni. Press, Cambaridge. 1965

[L–T] Levine, R.D. and Tribus, Myron, The Maximal Entropy Formalism, MIT Press, 1979

[M] Miller, R. W. Fact and Method, Princeton Uni. Press, 1987 Chap 6. The New Positivism, Chap. 7 Anti-Bayes

[N] Newman, James R. The World of Mathematics, Simon and Schuster, 1956

[P] Pinkham, Roger. On the Distribution of First Significant Digits, Annals of Math. Stat. 32(1961) pp. 1223–1230

[Pr] Primas, Hans. Chemistry, Quantum Mechanics and Reductionism, Springer-Verlag, Berlin (1983)

[R] Rand, One Million Random Digits and One Hundred Thousand Normal Deviates, Free Press, Glencoe, Ill.

[Ra] Ramakrishnan, S. and Sudderth, W.D. A Sequence of Coin Toss Variables for Which the Strong Law Fails. Am. Math. Monthly, Vol 95 (1988) No. 10, pp. 939–941

[St] Stoynov, J. Counterexmples, John Wiley and Sons, N.Y. 1987

[S] Sze'kely, G. Paradoxes in Probability Theory and Mathematical Statistics, D. Reidel Pub. Co. 1947

[T] Tribus, Myron. Thermostatics and Thermodynamics, D. van Nostrand Co. N.Y.C. 1961

[Wa] Wannier, G. H. Statistical Physics, Dover, 1966

[W] Waterhouse, W. C. Do Symmetric Problems Have Symmetric Solutions? Am. Math. Monthly, June–July 1983, p.378–387

[Z] Zipf, G. K. Human Behavior and the Principle of Least Effort, Addison-Wesley, 1949, reprint Hafner, 1965

Index

The Addison-Wesley **Advanced Book Program** would like to offer you the opportunity to learn about our new mathematics, statistics, and scientific computing titles in advance. To be placed on our mailing list and receive pre-publication notices and special offers, just **fill out this card completely** and return to us, postage paid. Thank you.

Title and Author of this book: **Date purchased:**

Name _____

Title _____

School/Company _____

Department _____

Street Address _____

City _____ State _____ Zip _____

Telephone/s(   ) _____ (   ) _____

Where did you buy/obtain this book?

☐ Bookstore   ☐ Mail Order   ☐ School (Required for Class)
☐ Campus Bookstore   ☐ Toll Free # to Publisher   ☐ Professional Meeting
☐ Other _____   ☐ Publisher's Representative

What professional mathematics and statistics associations are you an active member of?

☐ AMS (Amer Mathematical Society)   ☐ NCTM (Nat Counc Teachers Math)   ☐ SIAM (Soc Indust Applied Math)
☐ ASA (Amer Statistical Association)   ☐ ORSA (Oper Research Soc America)   ☐ AAAS (Amer Assoc for the Advancement of Science)
☐ MSA (Math Society of America)
☐ Other _____

Check your areas of interest.

(60) ✔**Mathematics/Statistics**

61 ☐ Advanced Calculus	69 ☐ Discrete Math	77 ☐ Operations Research
62 ☐ Algebra	70 ☐ Dynamical Systems	78 ☐ Optimization
63 ☐ Analysis	71 ☐ Geometry	79 ☐ Probability Theory
64 ☐ Applied Math	72 ☐ Logic/Probability	80 ☐ Statistical Modelling
65 ☐ Applied Statistics	73 ☐ Math-Biology	81 ☐ Stochastic Processes
66 ☐ Combinatorics	74 ☐ Math-Modelling	82 ☐ Time Series Analysis
67 ☐ Complex Variables	75 ☐ Math-Physics	83 ☐ Topology
68 ☐ Decision Theory	76 ☐ Number Theory	84 ☐ Other _____

Are you more interested in: ☐ pure math ☐ applied math?

Are you currently writing, or planning to write a textbook, research monograph, reference work, or create software in any of the above areas?
     ☐ Yes   ☐ No
     Area: _____

(If Yes) **Are you interested in discussing your project with us?**
     ☐ Yes   ☐ No

Mathematics/Statistics

No Postage
Necessary
if Mailed in the
United States

BUSINESS REPLY MAIL
FIRST CLASS PERMIT NO. 828   REDWOOD CITY, CA   94065

Postage will be paid by Addressee:

ADDISON-WESLEY
PUBLISHING COMPANY, INC.®

Advanced Book Program
350 Bridge Parkway, Suite 209
Redwood City, CA 94065-1522